EXPRESS
REVIEW GUIDES
Basic Math &
Pre-Algebra
LearningExpress
New York

Library of Congress Cataloging-in-Publication Data:
Express review guides. Basic math and pre-algebra.—1st ed.
p. cm.
ISBN: 978-1-57685-593-5
1. Mathematics—Outlines, syllabi, etc. 2. Algebra—Outlines, syllabi,
I. LearningExpress (Organization) II. Title: Basic math and pre-algebr
QA11.2.E97 2007
510—dc22

2007009013

Printed in the United States of America

9 8 7 6 5 4 3 2 1

First Edition

ISBN: 978-1-57685-593-5

For more information or to place an order, contact LearningExpress at
55 Broadway
8th Floor
New York, NY 10006

Or visit us at:
www.learnatest.com

Contents

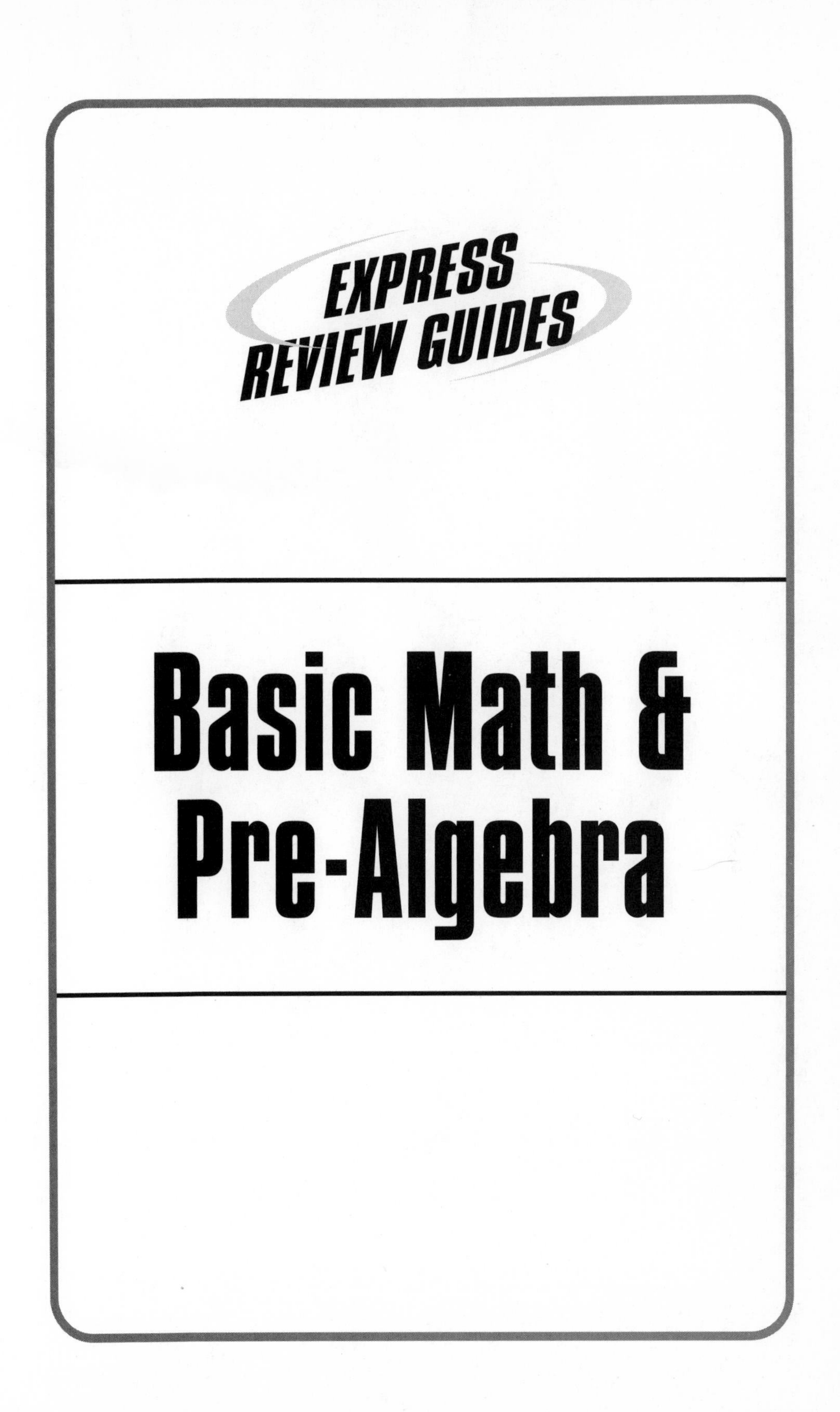

Basic Math & Pre-Algebra

1 Math? Why, oh Why?

Here's a common scenario that teachers run into. In every math classroom, without a doubt, a student will ask, "But why do we need to learn math?" Or, better yet, "Will we *ever* use this stuff in real life?"

Well, the answer is yes—I swear! If you cook a big meal for your family, you rely on math to follow recipes. Decorating your bedroom? You'll need math for figuring out what size bookshelf will fit in your room. Oh, and that motorcycle you want to buy? You're going to need math to figure out how much you can pay every month toward the total payment.

People have been using math for thousands of years, across various countries and continents. Whether you're sailing a boat off the coast of the Dominican Republic or building an apartment in Moscow, you're using math to get things done. You might be asking, how can math be so universal? Well, if you think about it, human beings didn't invent math concepts—we discovered them.

And then, there's algebra. Learning algebra is a little like learning another language. It's not exactly like English or Spanish or Japanese. Algebra is a simple language, used to create mathematical models of real-world situations and to help you with problems you aren't able to solve using plain old arithmetic. Rather than using words, algebra relies on numbers, symbols, and variables to make statements about things. And because algebra uses the same symbols as arithmetic for adding, subtracting, multiplying, and dividing, you're already familiar with the basic vocabulary!

People from different areas of the world and different backgrounds can "speak" algebra. If you are well versed in this language, you can use it to make important decisions and perform daily tasks with ease. Algebra can help you to shop intelligently on a budget, understand population growth, or even bet on the horse with the best chance of winning the race.

OKAY, BUT WHY *EXPRESS REVIEW GUIDES*?

If you're having trouble in math, this book can help you get on the right track. Even if you feel pretty confident about your math skills, you can use this book to help you review what you've already learned. Many study guides tell you how to improve your math—this book doesn't just *tell* you how to solve math problems, it *shows* you how. You'll find page after page of strategies that work, and you are never left stranded, wondering how to get the right answer to a problem. We'll show you all the steps to take so that you can successfully solve every single problem, and see the strategies at work.

Sometimes, math books assume you can follow explanations of difficult concepts even when they move quickly or leave out steps. That's not the case with this book. In each lesson, you'll find step-by-step strategies for tackling the different kinds of math problems. Then, you're given a chance to apply what you've learned by tackling practice problems. Answers to the practice problems are provided at the end of each section, so you can check your progress as you go along. This book is like your own personal math tutor!

THE GUTS OF THIS GUIDE

Okay, you've obviously cracked open the cover of this book if you're reading these words. But let's take a quick look at what is lurking in the other chapters. This book includes:

- a 50-question benchmark pretest to help you assess your knowledge of the concepts and skills in this guide
- brief, focused lessons covering essential basic math and pre-algebra topics, skills, and applications
- specific tips and strategies to use as you study

- a 50-question posttest followed by complete answer explanations to help you assess your progress

As you work through this book, you'll notice that the chapters are sprinkled with all kinds of helpful tips and icons. Look for these icons and the tips they provide. They include:

- *Fuel for Thought*: critical information and definitions that can help you learn more about a particular topic
- *Practice Lap*: quick practice exercises and activities to let you test your knowledge
- *Inside Track*: tips for reducing your study and practice time—without sacrificing accuracy
- *Caution!*: pitfalls to be on the lookout for
- *Pace Yourself*: try these extra activities for added practice

READY, SET, GO!

To best use this guide, you should start by taking the pretest. You'll test your math skills and see where you might need to focus your study.

Your performance on the pretest will tell you several important things. First, it will tell you how much you need to study. For example, if you got eight out of ten questions right (not counting lucky guesses!), you might need to brush up only on certain areas of knowledge. But if you got only five out of ten questions right, you will need a thorough review. Second, it can tell you what you know well (that is, which subjects you *don't* need to study). Third, you will determine which subjects you need to study in-depth and which subjects you simply need to refresh your knowledge.

REMEMBER . . .

THE PRETEST IS only practice. If you did not do as well as you anticipated, do not be alarmed and certainly do not despair. The purpose of the quiz is to help you focus your efforts so that you can *improve.* It is important to analyze your results carefully. Look beyond your score, and consider *why* you answered some questions incorrectly. Some questions to ask yourself when you review your wrong answers:

- Did you get the question wrong because the material was totally unfamiliar?
- Was the material familiar, but you were unable to come up with the right answer? In this case, when you read the right answer, it will often make perfect sense. You might even think, "I knew that!"
- Did you answer incorrectly because you read the question carelessly?

Next, look at the questions you got right and review how you came up with the right answer. Not all right answers are created equal.

- Did you simply know the right answer?
- Did you make an educated guess? An educated guess might indicate that you have some familiarity with the subject, but you probably need at least a quick review.
- Did you make a lucky guess? A lucky guess means that you don't know the material and you will need to learn it.

After the pretest, begin the lessons, study the example problems, and try the practice problems. Check your answers as you go along, so if you miss a question, you can study a little more before moving on to the next lesson.

After you've completed all the lessons in the book, try the posttest to see how much you've learned. You'll also be able to see any areas where you may need a little more practice. You can go back to the section that covers that skill for some more review and practice.

THE RIGHT TOOLS FOR THE JOB

BE SURE THAT you have all the supplies you need on hand before you sit down to study. To help make studying more pleasant, select supplies that you enjoy using. Here is a list of supplies that you will probably need:

- Notebook or legal pad
- Graph paper
- Pencils
- Pencil sharpener
- Highlighter
- Index or other note cards
- Paper clips or sticky note pads for marking pages
- Calendar or personal digital assistant (which you will use to keep track of your study plan)
- Calculator

As you probably realize, no book can possibly cover all of the skills and concepts you may be faced with. However, this book is not just about building a basic math and pre-algebra base, but also about building those essential skills that can help you solve unfamiliar and challenging questions. The basic math and pre-algebra topics and skills in this book have been carefully selected to represent a cross section of basic skills that can be applied in a more complex setting, as needed.

Pretest

By taking this pretest, you will get an idea of how much you already know and how much you need to learn about basic math and pre-algebra.

This pretest consists of 50 questions and should take about one hour to complete. You should not use a calculator when taking this pretest; however, you may use scratch paper for your calculations.

When you complete this pretest, compare your answers to the answer key at the end of the chapter. If your answers differ from the correct answers, use the explanations given to retrace your calculations. Along with each answer is a number that tells you which chapter of this book teaches you the math skills needed for that question.

1. $23 + 87 =$

2. $128 - 96 =$

3. What is the prime factorization of 54?

4. Reduce $\frac{21}{35}$.

5. Calculate $\frac{2}{5} + \frac{3}{5}$.

6. Calculate $\frac{2}{5}$ divided by $\frac{3}{5}$.

7. $\frac{4}{5} - \frac{2}{3} =$

8. Convert $5\frac{3}{8}$ to an improper fraction.

9. $2\frac{1}{8} + 5\frac{3}{8} =$

10. Calculate $3\frac{1}{4}$ multiplied by $5\frac{1}{2}$.

11. Arrange these numbers in order, from smallest to largest:
18; –3; 17.456; 17.320; 83; 8.3

12. Add the following numbers: 23; 2.3; 0.23; 2.056

13. Multiply 10.3 by 0.45.

14. 5.234 – 2.432 =

15. 3 is what percent of 4?

16. a is what percent of b?

17. During a period of one year, John earned $1,000 interest on a bank deposit at 5% interest. What was the principal at the beginning of the year?

18. Harry earned a 20% commission on a sale of $3,000. How much money did he earn on this commission?

19. Convert 35 to a percent value.

20. Find the absolute values of the following numbers: 7.2, –186, –(–1.3), and –(62).

21. 5 – (–8) =

22. $3 - 5 + 4 + 3 - 10 =$

23. $(-8) \cdot (-3) =$

DIRECTIONS: Use the following information to answer questions 24 through 27.

Here are four categories of numbers:

A. whole numbers
B. integers
C. rational numbers
D. real numbers

24. Which category or categories does 0 belong to?

25. Which category or categories does 3 belong to?

26. Which category or categories does $\sqrt{2}$ belong to?

27. Which category or categories does $3\frac{2}{3}$ belong to?

28. What is the value in pennies of n nickels?

29. Sally is s years old. How old was she 3 years ago?

30. Evaluate $4 + 5 \times 6$.

31. Evaluate $10a + b$ when $a = 5$ and $b = 6$.

32. Solve this equation for x: $3x + 8 = 23$.

33. For non-zero p, which is larger, p or $\frac{1}{p}$?

34. Which is larger, –3 or the reciprocal of –3?

35. Sally's take-home pay is $4,500. Before she gets her check, 25% of her pay is taken out. What is her salary before deductions?

36. A coat is discounted 10%, and then the store raises the price 10%. What is the net percent increase or decrease of the price?

37. Nine lions and tigers make up 60% of the animals in the lion house. How many other animals are in this facility?

38. $|13| =$

39. Which has a greater value, 0 or –1?

40. $-14 - (-11) =$

41. Reduce the ratio 80:5.

42. In a proportion, 3 is to 5 as 12 is to x. What is the value of x?

43. Which is longer, 3 inches or 7 centimeters?

44. How much is 12 feet, 4 inches plus 8 feet, 9 inches?

45. Approximately how many cubic feet are 3 quarts?

46. What is the area of each face of a cube that has a volume of 125 cm^3?

47. What is the perimeter of a rectangle that has a length of 4 inches and a width of 6 inches?

48. What is the volume of a sphere that has a radius of 10 inches?

49. Consider this data set: 1, 1, 2, 2, 2, 3. What is the (a) mean, (b) median, (c) mode, and (d) range?

50. If you added 10 to each of the 6 numbers in the data set of the previous problem, which of the statistics—(a) the mean, (b) the median, (c) the mode, and (d) the range—would change, and which would stay the same?

ANSWERS

1. 110; there is a carry from the units column to the tens column, and from the tens column to the hundreds column. For more help with this concept, see Chapter 3.

2. 32; for more help with this concept, see Chapter 3.

3. Is 54 divisible by 2? Yes, $\frac{54}{2} = 27$.
Is 27 divisible by 2? No. Is 27 divisible by 3? Yes, $\frac{27}{3} = 9$.
Is 9 divisible by 3? Yes, $9 = 3 \cdot 3$.
$54 = 2 \cdot 3 \cdot 3 \cdot 3$
Final answer: $2 \cdot 3^3$
For more help with this concept, see Chapter 3.

4. Express 21 and 35 in terms of their prime factors: $\frac{21}{35} = \frac{(3 \cdot 7)}{(5 \cdot 7)} = \frac{3}{5}$. For more help with this concept, see Chapter 4.

5. $\frac{2}{5} + \frac{3}{5} = \frac{5}{5} = 1$. For more help with this concept, see Chapter 4.

6. $\frac{2}{5} \div \frac{3}{5} = \frac{2}{5} \cdot \frac{5}{3} = \frac{2}{3}$. For more help with this concept, see Chapter 4.

7. $\frac{4}{5} - \frac{2}{3} = \frac{12}{15} - \frac{10}{15} = \frac{(12 - 10)}{15} = \frac{2}{15}$. For more help with this concept, see Chapter 4.

8. $5\frac{3}{8} = \frac{40}{8} + \frac{3}{8} = \frac{(40 + 3)}{8} = \frac{43}{8}$. For more help with this concept, see Chapter 4.

9. $2\frac{1}{8} + 5\frac{3}{8} = 7\frac{4}{8} = 7\frac{1}{2}$. For more help with this concept, see Chapter 4.

10. $3\frac{1}{4} \cdot 5\frac{1}{2} = \frac{13}{4} \cdot \frac{11}{2} = \frac{143}{8} = 17\frac{7}{8}$. For more help with this concept, see Chapter 4.

11. –3; 8.3; 17.320; 17.456; 18; 83. For more help with this concept, see Chapter 5.

12. Do zero-filling for neatness.

$$\begin{array}{r} 23.000 \\ 2.300 \\ 0.230 \\ +\ 2.056 \\ \hline 27.586 \end{array}$$

For more help with this concept, see Chapter 5.

13. There are three decimal places to take into account, so the answer is 4.635. For more help with this concept, see Chapter 5.

14.

$$\begin{array}{r} 5.234 \\ -\ 2.432 \\ \hline 2.802 \end{array}$$

For more help with this concept, see Chapter 5.

15. Change the question to "3 is what fraction of 4?" The answer to this question is $\frac{3}{4}$. Now multiply by 100% (which equals 1):
$(\frac{3}{4})(100)\% = [(\frac{3}{4})100]\% = [75]\% = 75\%$
For more help with this concept, see Chapter 6.

16. Change the question to "*a* is what fraction of *b*?" The answer is $\frac{a}{b}$. Now multiply by 100% (which equals 1): $(\frac{a}{b})100\% = [\frac{100a}{b}]\%$. For more help with this concept, see Chapter 6.

17. The fraction is 0.05. The special amount is $1,000.
Principal $= \frac{\$1{,}000}{0.05} = \$20{,}000$. For more help with this concept, see Chapter 6.

18. Harry earned 20% of $3,000, which is $0.20 \cdot \$3{,}000 = \$600$. For more help with this concept, see Chapter 6.

19. Multiply 35 by 100% (which equals 1): $35 \cdot 100 \cdot \% = (35 \bullet 100)\% =$ 3,500%. For more help with this concept, see Chapter 6.

20. The absolute value of 7.2 is 7.2; for –186, it is 186; for –(–1.3), it is 1.3; and for –(62), it is 62. For more help with this concept, see Chapter 7.

21. $5 - -8 = 5 + 8 = 13$. For more help with this concept, see Chapter 7.

22. $3 - 5 + 4 + 3 - 10 = -5$. Remember to change subtraction to adding the opposite. For more help with this concept, see Chapter 7.

23. $-8 \cdot -3 = 24$. The product of two negatives is positive. For more help with this concept, see Chapter 7.

24. Zero belongs to all the categories: A, B, C, and D. For more help with this concept, see Chapter 3.

25. The number 3 belongs to all the categories: A, B, C, and D. For more help with this concept, see Chapter 3.

26. $\sqrt{2}$ is irrational. It belongs only to category D. For more help with this concept, see Chapter 3.

27. This mixed number, $3\frac{2}{3}$, is a rational number. It belongs to categories C and D. For more help with this concept, see Chapter 3.

28. n nickels are worth $5n$ pennies. For more help with this concept, see Chapter 8.

29. Three years ago, Sally's age was $s - 3$. For more help with this concept, see Chapter 8.

30. Do the multiplication first: $4 + 5 \times 6 = 4 + 30 = 34$. For more help with this concept, see Chapter 8.

31. $10a + b = 10 \cdot 5 + 6 = 50 + 6 = 56$. For more help with this concept, see Chapter 8.

32. $3x + 8 = 23$.
$3x = 15$. Add–8 to both sides.
$x = 5$. Divide both sides by 3.
$3(5) + 8 = 23$. Plug in the 5 for x to check the result.
For more help with this concept, see Chapter 8.

33. People who answer, incorrectly, that p is larger than $\frac{1}{p}$ are thinking, "Maybe p is 3. Then 3 is larger than $\frac{1}{3}$." But what if p is $\frac{1}{3}$ and $\frac{1}{p}$ is 3? And what if p is -3? In fact, more information is needed about p. For more help with this concept, see Chapter 4.

34. Compare –3 with $-\frac{1}{3}$ on a number line. The $-\frac{1}{3}$ is closer to the origin, and thus is to the right of –3. The reciprocal of –3 is larger than –3. For more help with this concept, see Chapter 4.

35. Sally's $4,500 paycheck is 75% of her base pay; call it *B*. So 0.75*B* = $4,500. Solve for *B*: *B* = $6,000. For more help with this concept, see Chapter 6.

36. If the original price of the coat is *P*, the price after a 10% discount is 0.9*P*. The price is then raised by 10%. (Ten percent of what? Ten percent of the 0.9*P*.) The price is now 1.1 times the 0.9*P*, or 1.1 · 0.9*P* = 0.99*P*. The overall effect is a 1% discount. For more help with this concept, see Chapter 6.

37. Let *T* be the total number of animals in the lion house. Sixty percent of the total comprise the nine lions and tigers. Expressed algebraically, 0.6*T* = 9. Solving for *T* (divide both sides by 0.6), *T* = 15. So, the 15 animals comprise 9 lions and tigers and 6 other animals. For more help with this concept, see Chapter 6.

38. $|13| = 13$. For more help with this concept, see Chapter 7.

39. Zero is greater because on a number line, 0 is to the right of –1 and it has a greater value than –1. For more help with this concept, see Chapter 7.

40. $-14 - (-11) = -14 + 11 = -(14 - 11) = -3$. For more help with this concept, see Chapter 7.

41. Divide both parts of the ratio by 5. The ratio reduces to 16:1. For more help with this concept, see Chapter 10.

42. You can write the proportion in fraction form as $\frac{3}{5} = \frac{12}{x}$. Cross multiply and solve for *x*: $3x = 60$; $x = 20$. The proportion is 3 is to 5 as 12 is to 20. For more help with this concept, see Chapter 10.

43. The answer is 3 inches. There are 2.54 centimeters in one inch. Therefore, 3 inches are equal to 2.54×3 centimeters, or 7.62 centimeters. For more help with this concept, see Chapter 9.

44. The feet add up to 20 feet (12 + 8 = 20), and the inches add up to 13 inches (4 + 9 = 13). Convert the 13 inches to 1 foot and 1 inch. The result is 21 feet, 1 inch. For more help with this concept, see Chapter 9.

45. Because 1 gallon = 4 quarts, 3 quarts = 0.75 gallon. One gallon is 0.1337 cubic foot. Therefore, 0.75 gallon is (0.1337)(0.75), which equals 0.100275, or approximately 0.1 cubic foot. For more help with this concept, see Chapter 9.

46. The volume of a cube is equal to s^3, where s is the length of one side of the cube: $s^3 = 125$ and $s = 5$. The length of one side of the cube is 5 centimeters. Each face of the cube is a square with area $A = s^2$. For $s = 5$ cm, $s^2 = 25$ cm^2. For more help with this concept, see Chapter 9.

47. The formula for perimeter of a rectangle is $P = 2l + 2w$, where l and w are length and width. Plugging in the given values, we have $P = 2(4$ inches$) + 2(6$ inches$) = 8$ inches $+ 12$ inches $= 20$ inches. For more help with this concept, see Chapter 9.

48. The formula for the volume of a sphere is $V = (\frac{4}{3})\pi r^3$. $V = \frac{4}{3}\pi(10)^3$, or $V = \frac{4}{3}\pi(1{,}000)$. $V = \frac{4{,}000\pi}{3}$. For more help with this concept, see Chapter 9.

49. The sum is 11 and the number of numbers is 6. So (a) the mean is $\frac{11}{6} = 1.83$. The middle numbers (the third and fourth numbers) are both 2, so (b) the median is 2. The 2 appears three times, which is more times than any of the other numbers appear, so (c) the mode is 2. The largest number is 3 and the smallest is 1, so the range is 2 (3 – 1 = 2). For more help with this concept, see Chapter 10.

50. The three measures of where the middle is (the mean, the median, and the mode) would each increase by 10. The measure of scatter (the range) would remain unchanged. For more help with this concept, see Chapter 10.

3 Numbers, Numbers Everywhere!

WHAT'S AROUND THE BEND

- Counting Numbers
- Integers
- Rational and Real Numbers
- Place Value
- Addition
- Subtraction
- Multiplication
- Division
- Prime Numbers
- Factoring Whole Numbers

Numbers, numbers, numbers. Before we begin to explore mathematical concepts and properties, let's discuss number terminology. The counting numbers, 0, 1, 2, 3, 4, 5, 6, and so on, are also known as the **whole numbers**. No fractions or decimals are allowed in the world of whole numbers. *What a wonderful world*, you say. No pesky fractions and bothersome decimals.

But, as we leave the tranquil world of whole numbers and enter into the realm of **integers**, we are still free of fractions and decimals, but subjected to the negative counterparts of all those whole numbers that we hold so dear. The set of integers would be $\{\ldots -3, -2, -1, 0, 1, 2, 3 \ldots\}$.

The **real numbers** include any number that you can think of that is not imaginary. You may have seen the imaginary number ***i***, or maybe you haven't (maybe you only *thought* you did). The point is, you don't have to worry about it. Just know that imaginary numbers are not allowed in the set of real numbers. No pink elephants either! Numbers that are included in the real numbers are fractions, decimals, zero, π, negatives, and positives of many varieties.

So where do irrational and rational numbers fit into all this? Here's how it works. The first five letters of RATIONAL are R-A-T-I-O. **Rational numbers** can be represented as a *ratio* of two integers. In other words, a rational number can be written as a decimal that either *ends* or *repeats*. **Irrational numbers** can't be represented as a ratio, because their decimal extensions go on and on forever without repeating. π is the most famous irrational number. Other irrational numbers are $\sqrt{2}$ and $\sqrt{11}$.

So, you probably think you know the deal with whole numbers, right? They are the numbers 0, 1, 2, 3, and so forth. Yes, you *already* know that they go on forever and that there is no such thing as the largest whole number.

But in order to conquer basic math and pre-algebra, you need to really understand how whole numbers work with the basic operations: addition, subtraction, multiplication, and division. And then there are the issues of place value, prime numbers, factoring, and . . .

Okay, ready to reintroduce yourself to whole numbers? They really are quite fascinating once you get to know them.

HOW TO BUILD LARGE NUMBERS OUT OF SMALL NUMBERS

You can write thousands of words using only the 26 letters of the English language alphabet. Think about the words you can create with just the letters M, A, T, and H. Can you also be this clever with whole numbers?

FUEL FOR THOUGHT

***NUMBER* IS A** very general term that includes whole numbers, fractions (e.g., $\frac{2}{3}$), decimals (e.g., 24.35), and negative numbers (e.g., −24).

Digit refers to the ten special numbers 0, 1, 2, 3, 4, 5, 6, 7, 8, and 9. The number 36 is a two-digit number.

Yes, you can be extremely creative with numbers. Using only ten basic numbers, you can write numbers as large as you like. Use these basic numbers—0, 1, 2, 3, 4, 5, 6, 7, 8, and 9—to write at least ten different number combinations on the lines below. Make them any length you want—just be sure to fit them on the page!

________	________	________	________	________	________
________	________	________	________	________	________
________	________	________	________	________	________
________	________	________	________	________	________
________	________	________	________	________	________
________	________	________	________	________	________
________	________	________	________	________	________
________	________	________	________	________	________
________	________	________	________	________	________
________	________	________	________	________	________

Okay, now you see at least ten different combinations for the basic numbers, but there are actually infinite combinations! No, we weren't going to make you try to list them all. We aren't that cruel! (We hear your sigh of relief.)

These ten basic numbers are called **digits** and they are very useful. If you didn't have them, imagine how you would have to express the number of days in a week. You would have to say something like, "There are 1,111,111 days in a week." Then, maybe you would eventually develop shortcuts, like using *v* to mean five: "There are vii days in a week." This was the beginning of the Roman numeral system, which is widely used today for lists that are not likely to go beyond about a dozen items.

Today, you have a much better system than Roman numerals. You have the place-value system, and with the place-value system, you can efficiently write numbers as large as you like.

In the place-value system, you can indicate the number of days in a year as 365. The 5 stands for 5 days; the 6 in the second place (counting from the right) stands for 6 times 10, or 60 days; and the 3 in the third place stands for 3 times 100, or 300 days. The values of the digits get ten times bigger each time you move to the left.

When you write 365, you are adding 300, 60, and 5 (300 + 60 + 5). If you put these numbers under each other, you can clearly see that they add up to 365.

$$\begin{array}{r} 300 \\ 60 \\ +\ \ 5 \\ \hline 365 \end{array}$$

HOW TO COUNT ANY NUMBER OF THINGS, NO MATTER HOW LARGE, WITH DIGITS

In any number, when you add on a digit, the original number grows by a factor of ten. It works like this:

The 9 in 9,321 signifies 9 thousand.
The 9 in 94,321 signifies 90 thousand.
The 9 in 954,321 signifies 900 thousand.
The 9 in 9,654,543 signifies 9 million.
The 9 in 97,654,321 signifies 90 million.

But there is an even more powerful way to move to larger numbers. Instead of increasing by tens, you can increase by factors of a thousand. You can move forward by groups of three digits, called **periods**.

FUEL FOR THOUGHT

A **period** is a group of three digits occurring in a number that represents 1,000 or larger. Here are the periods:

units
thousands
millions
billions
trillions
quadrillions
quintillions
sextillions
septillions
octillions
nonillions

53 thousand is written as 53,000.
53 million is written as 53,000,000.
53 billion is written as 53,000,000,000.
53 trillion is written as 53,000,000,000,000.
53 quadrillion is written as 53,000,000,000,000,000.
53 quintillion is written as 53,000,000,000,000,000,000.
53 sextillion is written as 53,000,000,000,000,000,000,000.
53 septillion is written as 53,000,000,000,000,000,000,000,000.
53 octillion is written as 53,000,000,000,000,000,000,000,000,000.
53 nonillion is written as 53,000,000,000,000,000,000,000,000,000,000.

Let's look at the number 557,987,654,321. What place value does each period represent?

The first (right-most) group of three digits, 321, signifies 321 units.
The second group of three digits, 654, signifies 654 thousand.
The third group of three digits, 987, signifies 987 million.
The fourth group of three digits, 557, signifies 557 billion.

Okay, you get the point. The whole number 557,987,654,321 is read as 557 billion, 987 million, 654 thousand, 321. The place-value system solves the problem of writing large numbers.

PRACTICE LAP

DIRECTIONS: Use scratch paper to solve the following problems. You can check your answers at the end of this chapter.

1. Write the number 1 billion, 234 million, 567 thousand, 890 by using only digits.
2. Write the following number, using period names: 9,876,543,210.
3. What is the value of the 7 in the number 9,876,543,210?

TO KNOW ADDITION IS TO LOVE ADDITION

Addition is simply the totaling of a column (or columns) of numbers. Addition answers the following question: How much is this number plus that number? You can use addition to figure out how many DVDs you own, or how much money you spent on those DVDs.

Example

Oliver has 61 graphic novels and Jennifer has 52 graphic novels. Oliver's sister Lena has an additional 37 graphic novels. They decide to store all their graphic novels on Oliver's bookshelf. How many graphic novels are on the bookshelf?

Start by adding the columns of numbers.

$$\begin{array}{r} 61 \\ 52 \\ +\ 37 \\ \hline \end{array}$$

Hopefully, you figured out that they have put 150 graphic novels on Oliver's bookshelf.

INSIDE TRACK

HOW TO CHECK YOUR ANSWERS

WHEN YOU ADD columns of numbers, how do you know if you came up with the right answer? One way to check, or proof, your answer is to add the figures from the bottom to the top. In the example you just saw, start with 7 + 2 in the right (ones) column and work your way up. Then, carry the 1 into the second column and say 1 + 3 + 5 + 6 and work your way up again. Your answer should still come out to 150.

Single-Digit Adding Facts

One important step to mastering addition is to know how to flash on (that is, know by heart) the single-digit adding facts, like 5 + 2 = 7 or 6 + 8 = 14.

In turn, the way to learn to flash on these facts is first to learn how to figure them out. If you don't flash on the single-digit adding facts, you will not be able to quickly and accurately add a column of multi-digit numbers. You might always feel a little uneasy around numbers.

On the other hand, if you have learned to flash on the single-digit adding facts, you will be able to manage any adding problem easily. Following is a chart of the single-digit adding facts. Once you are familiar with these facts, you will be able to add more comfortably.

ZERO

$0 + 0 = 0$ | $0 + 1 = 1$ | $0 + 2 = 2$ | $0 + 3 = 3$ | $0 + 4 = 4$
$0 + 5 = 5$ | $0 + 6 = 6$ | $0 + 7 = 7$ | $0 + 8 = 8$ | $0 + 9 = 9$

ONE

$1 + 0 = 1$ | $1 + 1 = 2$ | $1 + 2 = 3$ | $1 + 3 = 4$ | $1 + 4 = 5$
$1 + 5 = 6$ | $1 + 6 = 7$ | $1 + 7 = 8$ | $1 + 8 = 9$ | $1 + 9 = 10$

TWO

$2 + 0 = 2$ | $2 + 1 = 3$ | $2 + 2 = 4$ | $2 + 3 = 5$ | $2 + 4 = 6$
$2 + 5 = 7$ | $2 + 6 = 8$ | $2 + 7 = 9$ | $2 + 8 = 10$ | $2 + 9 = 11$

THREE

$3 + 0 = 3$ | $3 + 1 = 4$ | $3 + 2 = 5$ | $3 + 3 = 6$ | $3 + 4 = 7$
$3 + 5 = 8$ | $3 + 6 = 9$ | $3 + 7 = 10$ | $3 + 8 = 11$ | $3 + 9 = 12$

FOUR

$4 + 0 = 4$ | $4 + 1 = 5$ | $4 + 2 = 6$ | $4 + 3 = 7$ | $4 + 4 = 8$
$4 + 5 = 9$ | $4 + 6 = 10$ | $4 + 7 = 11$ | $4 + 8 = 12$ | $4 + 9 = 13$

FIVE

$5 + 0 = 5$ | $5 + 1 = 6$ | $5 + 2 = 7$ | $5 + 3 = 8$ | $5 + 4 = 9$
$5 + 5 = 10$ | $5 + 6 = 11$ | $5 + 7 = 12$ | $5 + 8 = 13$ | $5 + 9 = 14$

SIX

$6 + 0 = 6$ | $6 + 1 = 7$ | $6 + 2 = 8$ | $6 + 3 = 9$ | $6 + 4 = 10$
$6 + 5 = 11$ | $6 + 6 = 12$ | $6 + 7 = 13$ | $6 + 8 = 14$ | $6 + 9 = 15$

SEVEN

$7 + 0 = 7$ | $7 + 1 = 8$ | $7 + 2 = 9$ | $7 + 3 = 10$ | $7 + 4 = 11$
$7 + 5 = 12$ | $7 + 6 = 13$ | $7 + 7 = 14$ | $7 + 8 = 15$ | $7 + 9 = 16$

EIGHT

$8 + 0 = 8$ | $8 + 1 = 9$ | $8 + 2 = 10$ | $8 + 3 = 11$ | $8 + 4 = 12$
$8 + 5 = 13$ | $8 + 6 = 14$ | $8 + 7 = 15$ | $8 + 8 = 16$ | $8 + 9 = 17$

NINE

$9 + 0 = 9$ | $9 + 1 = 10$ | $9 + 2 = 11$ | $9 + 3 = 12$ | $9 + 4 = 13$
$9 + 5 = 14$ | $9 + 6 = 15$ | $9 + 7 = 16$ | $9 + 8 = 17$ | $9 + 9 = 18$

INSIDE TRACK

WHEN WORKING WITH single-digit addition facts, there are several rules to keep in mind.

Zero plus a whole number always equals that number.

$0 + 8 = 8$

$0 + 9 = 9$

Adding one to a whole number moves you to the next whole number.

$1 + 8 = 9$

$1 + 9 = 10$

Two plus a number is like counting twice from that number. You will always end up on the next odd or even number, depending on whether the original number was odd or even.

$8 + 2 = 10$ (10 is the next even number after 8.)

$3 + 2 = 5$ (5 is the next odd number after 3.)

It is sometimes easier to add using tens, when possible. For example, when adding 9 plus a number, bump the 9 up one, to 10, and bump the other number down one.

$9 + 7 = 10 + 6 = 16$

When adding 8 plus a number, bump the 8 up two, to 10, and bump the other number down two.

$8 + 7 = 10 + 5 = 15$

Whenever it is convenient to bump one number up to 10 and the other number down, do it.

$7 + 6 = 10 + 3 = 13$

You know that 5 plus 5 equals 10. You can use this basic addition fact to add other numbers by learning what 6, 7, 8, and 9 are in terms of 5 plus how many. No matter how long it takes, memorize these four facts:

$6 = 5 + 1$
$7 = 5 + 2$
$8 = 5 + 3$
$9 = 5 + 4$

Knowing these four facts, you have now conquered the whole lower right part of the addition table. It's that easy.

Example

6 + 8 = ?

Think of 6 as 5 + 1 and think of 8 as 5 + 3. Now, you have (5 + 1) + (5 + 3). Rearrange the terms to group the 5s together: 5 + 5 = 10 and 1 + 3 = 4; 10 + 4 = 14, so 6 + 8 = 14.

Adding Columns of Digits

Add this column of digits:

$$\begin{array}{r} 8 \\ 5 \\ 2 \\ 7 \\ 5 \\ 6 \\ 2 \\ 4 \\ 1 \\ +\ 6 \\ \hline \end{array}$$

If you flash on all your single-digit adding facts, then you can add this column in order, quickly, as follows: 8, 13 (5 + 8), 15 (13 + 2), 22 (15 + 7), 27 (22 + 5), 33 (27 + 6), 35 (33 + 2), 39 (35 + 4), 40 (39 + 1), 46 (40 + 6). As a check, add the numbers from the bottom up: 6, 7, 11, 13, 19, 24, 31, 33, 38, 46.

But suppose you don't know how to flash. Suppose that, like many people, the only addition facts you know well are the ones that add to ten, like 8 + 2 = 10 and 6 + 4 = 10. Then, you can use a trick to help you add better—look for combinations of tens. Tens are easy to add. In our example, seeing the 8 at the top of the list, you look for and find a 2. That makes your first ten. There are two 5s, and that makes for another ten, so you have 10 + 10 = 20. You will also find ten in 7 + 2 + 1 and another ten in 6 + 4. That makes 40, and there is a 6 left over, for a grand total of 46.

Adding Multi-Digit Numbers

Let's say you've been asked to add 42 and 56. How do you do this? You could start at 42 and count 56 times. A different way is to start at 56, and then you have to count only 42 times. However, both these methods would take a long time.

Fortunately, there is a faster way—using the place-value system. Look at the number 42 and you see 4 tens and 2 ones. In 56, you'll see 5 tens and 6 ones.

Using the place-value system, you can add 4 tens and 5 tens to get 9 tens, and 2 units and 6 units to get 8 units; 9 tens and 8 units is written as 98.

$$\begin{array}{r} 42 \\ +\ 56 \\ \hline 98 \end{array}$$

Okay, now how much is 57 + 85? This example has a little wrinkle that the first one doesn't have. You'll have to "carry."

Write the numbers one under the other, as before. Then, you have the units (7 + 5 = 12) and the tens digits (5 + 8 = 13).

When the sum of a column gets to be ten or more than ten, you need to *carry.* In the example above, the 7 and the 5 in the units digits result in 12. We write down the 2 and carry the 1 over to the next column, where the digit 1 signifies 1 ten.

If the numbers in the hundreds column add to a two-digit number like 12, the 2 signifies 2 hundreds and the 1 signifies 10 hundreds, which is 1 thousand. (The thousands column is right there, the column to the left of the hundreds column.)

$$\begin{array}{r} {}^{1} \\ 57 \\ +\ 85 \\ \hline 142 \end{array}$$

Now try your hand at adding three columns of numbers.

$$
\begin{array}{r}
196 \\
312 \\
604 \\
537 \\
578 \\
943 \\
+\ 725 \\
\hline
\end{array}
$$

Did you get the correct answer? You'll know for sure if you proofed it. If you haven't, then go back right now and check your work.

PACE YOURSELF

HOW TO "WRITE" A DIGIT ON ONE HAND

LET EACH FINGER represent one, and your thumb represent five. Then, you can represent the ten digits like this:

0 no fingers
1 finger
2 finger finger
3 finger finger finger
4 finger finger finger finger
5 thumb
6 thumb finger
7 thumb finger finger
8 thumb finger finger finger
9 thumb finger finger finger finger

$$
\begin{array}{r}
{}^{2\,3}\\
196\\
312\\
604\\
537\\
578\\
943\\
+\ 725\\
\hline
3{,}895
\end{array}
$$

Did you get it right? Did you come up with 3,895? If you still feel a little rusty, then what you need is more practice, and the following exercises will help.

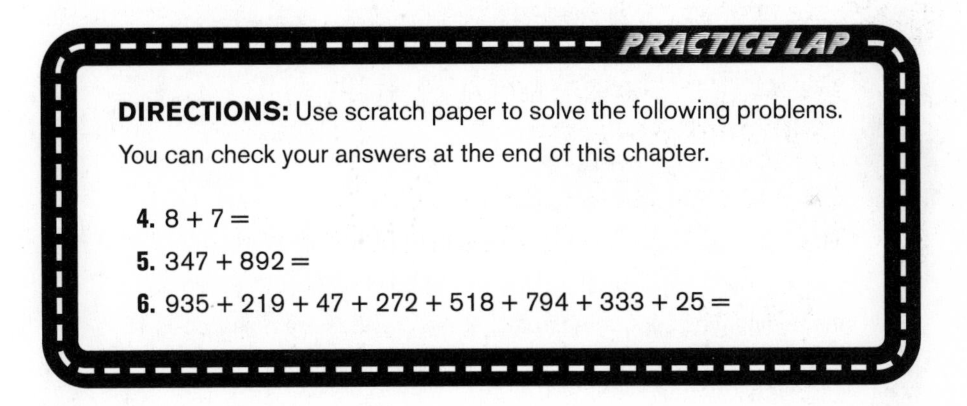

PRACTICE LAP

DIRECTIONS: Use scratch paper to solve the following problems. You can check your answers at the end of this chapter.

4. 8 + 7 =

5. 347 + 892 =

6. 935 + 219 + 47 + 272 + 518 + 794 + 333 + 25 =

TACKLING SUBTRACTION

Subtraction is the mathematical opposite of addition. Instead of combining one number with another, you take one away from another.

Let's start off by working out some basic subtraction problems. These problems are simple because you don't have to borrow or cancel any numbers. Try working out these problems on your own before you scroll down to the answers.

Examples

$$\begin{array}{r} 68 \\ -\ 53 \\ \hline \end{array} \qquad \begin{array}{r} 77 \\ -\ 36 \\ \hline \end{array} \qquad \begin{array}{r} 94 \\ -\ 41 \\ \hline \end{array}$$

FUEL FOR THOUGHT

THE RESULT OF subtracting 5 from 8, which is 3, is called the **difference** between 8 and 5.

When you calculate 8 minus 5, the 8 is called the **minuend** and the 5 is called the **subtrahend**.

$$\begin{array}{r} 68 \\ -\ 53 \\ \hline 15 \end{array} \qquad \begin{array}{r} 77 \\ -\ 36 \\ \hline 41 \end{array} \qquad \begin{array}{r} 94 \\ -\ 41 \\ \hline 53 \end{array}$$

INSIDE TRACK

THERE'S A GREAT way to check or proof your answers. Just add your answer to the number you subtracted and see if they add up to the number you subtracted from. For the previous examples, does 15 + 53 = 68? Does 41 + 36 = 77? Does 53 + 41 = 94?

Multi-Digit Subtraction

Now, I'll add a wrinkle. You're going to need to "borrow." Are you ready?

Example

$$\begin{array}{r} 54 \\ -\ 49 \\ \hline \end{array}$$

You need to subtract 9 from 4. Well, that's pretty hard to do. So, you need to make the 4 into 14 by borrowing 1 from the 5 of 54. What you're doing here is converting one ten (a 1 in the tens place) to ten ones. Okay, so 14 – 9 is 5. Because you borrowed 1 from the 5, the 5 is now 4. And 4 – 4 is 0. So, 54 – 49 = 5.

Now, we'll take subtraction one step further. Do you know how to do the two-step dance? If not, no worries. Now you're going to be doing the subtraction three-step, or at least subtracting with three-digit numbers.

Examples

$$\begin{array}{r} 532 \\ -\,149 \\ \hline \end{array} \qquad \begin{array}{r} 903 \\ -\,616 \\ \hline \end{array} \qquad \begin{array}{r} 682 \\ -\,493 \\ \hline \end{array}$$

Solutions

$$\begin{array}{r} {}^{4\ 12\ 12} \\ \not{5}\,\not{3}\,\not{2} \\ -\,1\,4\,9 \\ \hline 3\,8\,3 \end{array} \qquad \begin{array}{r} {}^{8\ 9\ 13} \\ \not{9}\,\not{0}\,\not{3} \\ -\,616 \\ \hline 287 \end{array} \qquad \begin{array}{r} {}^{5\ 17\ 12} \\ \not{6}\,\not{8}\,\not{2} \\ -\,4\,9\,3 \\ \hline 1\,8\,9 \end{array}$$

PRACTICE LAP

DIRECTIONS: Use scratch paper to solve the following problems. You can check your answers at the end of this chapter.

7. 12 – 7 =
8. 840 – 162 =
9. A school bus has 43 children on board. They are being taken to four schools; 12 children get off at the first school, 7 at the second school, and 15 at the third school. How many children are left on the bus to go to the fourth school?

WELCOME TO THE STAGE . . . MULTIPLICATION!

Multiplication is really just another form of addition. For instance, how much is 5 × 4? You might know it's 20 because you searched your memory for that multiplication fact. There's nothing wrong with that. As long as you can remember what the answer is from the multiplication table, you're all right.

Another way to calculate 5 × 4 is to add them: 4 + 4 + 4 + 4 + 4 = 20.

You use multiplication in place of addition, because it's shorter. Suppose you had to multiply 395 × 438. If you set this up as an addition problem, you'd be working at it for a couple of hours.

Do you know the multiplication table? You might know most of these facts by heart, but many people have become so dependent on their calculators that they've forgotten a few multiplication problems—like 9 × 6 or 8 × 7.

Multiplication is basic in understanding math. And to really know how to multiply, you need to know the entire multiplication table by memory.

Test yourself. First, fill in the answers to the multiplication problems in the table that follows. Then check your work against the numbers shown in the completed multiplication table on page 38. If they match, then you have a good grasp of your multiplication facts. But if you missed a few, then you need to practice those until you've committed them to memory.

X	0	1	2	3	4	5	6	7	8	9	10
0											
1											
2											
3											
4											
5											
6											
7											
8											
9											
10											

FUEL FOR THOUGHT

8 × 5 IS THE SAME AS 5 × 8

Here is a diagram of 8 rows, each one having 5 bagels.

BBBBB
BBBBB
BBBBB
BBBBB
BBBBB
BBBBB
BBBBB
BBBBB

If you look at the diagram differently, you can see 5 columns, each one having 8 bagels.

These two ways of looking at the diagram demonstrate that 8 × 5 is the same as 5 × 8. This concept is true of all the multiplying facts.

Let's look into the realm of long multiplication. Long multiplication is just multiplication combined with addition.

Example

$$\begin{array}{r} 46 \\ \times\ 37 \\ \hline \end{array}$$

First, multiply 6 by 7, which gives us 42. Write down the 2 and carry the 4:

$$\begin{array}{r} {}^{4} \\ 46 \\ \times\ 37 \\ \hline 2 \end{array}$$

Now, multiply 4 by 7, which gives you 28. Add the 4 you carried to 28 and write down 32:

$$\begin{array}{r} \overset{4}{4}6 \\ \times\ 37 \\ \hline 322 \end{array}$$

Next, multiply 6 by 3, which should yield 18. Write down the 8 and carry the 1:

$$\begin{array}{r} \overset{1}{4}6 \\ \times\ 37 \\ \hline 322 \\ 8 \end{array}$$

Now, calculate 4 $\times$ 3, giving you 12. Add the 1 you carried to 12, and write down 13:

$$\begin{array}{r} \overset{1}{4}6 \\ \times\ 37 \\ \hline 322 \\ 138 \end{array}$$

Then, you can add your columns:

$$\begin{array}{r} \overset{1}{4}6 \\ \times\ 37 \\ \hline 322 \\ 138 \\ \hline 1{,}702 \end{array}$$

INSIDE TRACK

TO PROVE YOUR multiplication, just reverse the numbers you are multiplying. For the preceding example, check your work this way:

$$\begin{array}{r} 37 \\ \times\ 46 \\ \hline 222 \\ 148 \\ \hline 1{,}702 \end{array}$$

You should still end up with the same final product.

Try another problem just to test your skills.

Example

$$\begin{array}{r} 89 \\ \times\ 57 \\ \hline \end{array}$$

Well, 7 × 9 = 63. Write down the 3 and carry the 6.

$$\begin{array}{r} \overset{6}{8}9 \\ \times\ 57 \\ \hline 3 \end{array}$$

You know that 7 × 8 = 56 and 56 + 6 = 62. Write down 62.

$$\begin{array}{r} 89 \\ \times\ 57 \\ \hline 623 \end{array}$$

Next, 5 × 9 = 45, so write down the 5 and carry the 4.

$$\begin{array}{r} \overset{4}{8}9 \\ \times\ 57 \\ \hline 623 \\ 5 \end{array}$$

Then, 5 × 8 = 40 and 40 + 4 = 44. Write down 44. Then, add the two columns.

$$\begin{array}{r} 89 \\ \times\ 57 \\ \hline 623 \\ 445 \\ \hline 5{,}073 \end{array}$$

PACE YOURSELF

FLASH CARDS

A good way to drill basic multiplication facts is to create flash cards. Using 3-by-5 index cards, you can create a useful set of flash cards for studying on the go. For example, write "6 × 8" and "8 × 6" on one side of an index card, and "48" on the other side. You can study either side of the flash card and try to say what's on the other side. This will remind you that you can get a product, like 48, by multiplying different pairs of numbers.

INSIDE TRACK

HERE ARE SOME single-digit multiplication facts:

Zero times any number is zero.

$0 \times 8 = 0$

One times any number is that number.

$1 \times 8 = 8$

Two times any number is that number plus itself.

$2 \times 8 = 8 + 8 = 16$

Three times any number is the number plus the double of the number.

$3 \times 8 = 8 + 2 \times 8 = 8 + 16 = 24$

Four times any number can be calculated by doubling the number, and then doubling that result (because $4 = 2 \times 2$). To calculate 4×7, double the 7 (14) and then double the 14 (28).

Five times any even number is like taking half the number and then multiplying by ten. For 5×8, take half of 8 (4) and multiply by 10 (40).

Six times any number can be calculated by tripling the number and then doubling that result (because $6 = 3 \times 2$). For 6×8, triple the 8 (24), and then double 24 (48).

Eight times any number can be calculated by doubling the number, and then doubling that result, and then doubling that last result once again. To calculate 8×6, double the 6 (12), then double the 12 (24), and then double the 24 (48).

Nine times any number can be calculated by tripling the number and then tripling that result. For 9×9, triple 9 (27) and then triple 27 (81).

X	0	1	2	3	4	5	6	7	8	9	10
0	0	0	0	0	0	0	0	0	0	0	0
1	0	1	2	3	4	5	6	7	8	9	10
2	0	2	4	6	8	10	12	14	16	18	20
3	0	3	6	9	12	15	18	21	24	27	30
4	0	4	8	12	16	20	24	28	32	36	40
5	0	5	10	15	20	25	30	35	40	45	50
6	0	6	12	18	24	30	36	42	48	54	60
7	0	7	14	21	28	35	42	49	56	63	70
8	0	8	16	24	32	40	48	56	64	72	80
9	0	9	18	27	36	45	54	63	72	81	90
10	0	10	20	30	40	50	60	70	80	90	100

PRACTICE LAP

DIRECTIONS: Use scratch paper to solve the following problems. You can check your answers at the end of this chapter.

10. Express 5 + 5 + 5 + 5 + 5 + 5 + 5 + 5 as a multiplication problem and give the result.

11. $9 \times 7 =$

12. $348 \times 624 =$

13. Mary has 17 marbles in each of 7 bags. How many marbles does she have all together?

AND FINALLY . . . DIVISION

As you'll see, division is the opposite of multiplication. So you really must know the multiplication table to do division problems. Let's look at a short division problem.

$$7\overline{)2,114}$$

You can divide 7 into 21 to get 3.

$$\begin{array}{r} 3 \\ 7\overline{)2,114} \end{array}$$

Then, try to divide 7 into 1. Because 7 is larger than 1, it doesn't fit. So we write 0 over the 1:

$$\begin{array}{r} 30 \\ 7\overline{)2,114} \end{array}$$

How many times does 7 go into 14?

$$\begin{array}{r} 302 \\ 7\overline{)2,114} \end{array}$$

You can check your answer using multiplication: $302 \times 7 = 2,114$.

This example came out even, but sometimes, there's a remainder. That's the case in the examples that follow.

Examples

$$\begin{array}{r} 45 \text{ R8} \\ 9\overline{)413} \end{array}$$

$$\begin{array}{r} 40 \text{ R1} \\ 8\overline{)321} \end{array}$$

Long Division

Long division is carried out in two steps:

1. trial and error
2. multiplication

The process of long division is identical to short division, but it involves a lot more calculation. That's why it's so important to have memorized the multiplication table.

Example

$$37\overline{)596}$$

How many times does 37 go into 59? Just once. So, put a 1 directly above the 9 and write 37 directly below 59.

$$\begin{array}{r} 1 \\ 37\overline{)596} \\ \underline{-37} \\ 22 \end{array}$$

Then, you subtract 37 from 59, leaving you with 22. Next, bring down the 6, giving you 226. How many times does 37 go into 226? You need to do this by trial and error. You finally come up with 6, because 6 × 37 = 222.

The **dividend** is the number being divided; the **divisor** is the number that divides the dividend. The **quotient** is the result.

In the division problem 12 ÷ 4, 12 is the dividend and 4 is the divisor. The quotient is 3.

In the division problem 13 ÷ 4, 3 is the quotient and there is a remainder of 1.

```
    16
37)596
  - 37
   226
 - 222
```

When you subtract 222 from 226, you are left with 4, which is the remainder.

```
    16 R4
37)596
  -37
   226
 - 222
     4
```

The proper notation for the answer is 16 R4. To check this answer, multiply 16 and 37 and add 4. Did you get 596?

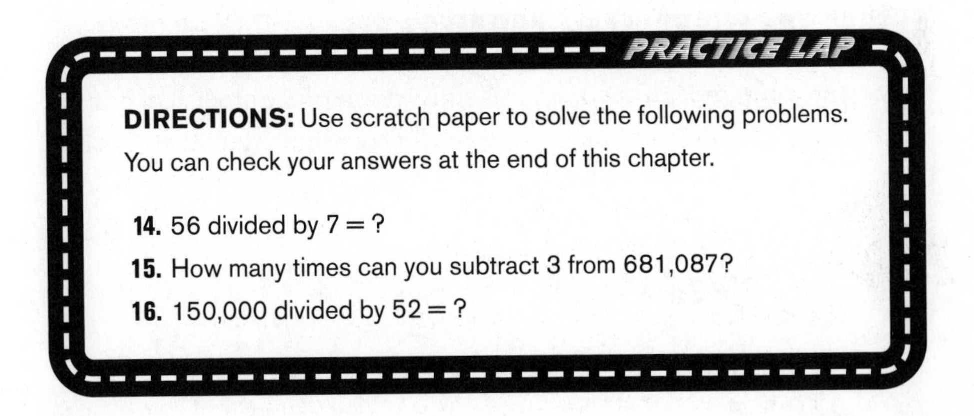

PRACTICE LAP

DIRECTIONS: Use scratch paper to solve the following problems. You can check your answers at the end of this chapter.

14. 56 divided by 7 = ?

15. How many times can you subtract 3 from 681,087?

16. 150,000 divided by 52 = ?

A GLIMPSE OF PRIME NUMBERS

Prime numbers are whole numbers greater than 1, whose only factors are 1 and themselves.

You can easily identify prime numbers. For example, look at the number 10. The factors of 10 are 2 and 5. So, 10 is not prime. Ten is a **composite number**, because it is "composed" (in a multiplicative sense) of 2 and 5.

On the other hand, 11 is a prime number. Its only factors are 1 and 11.

Every whole number greater than 1 is either prime or composite. Here are the first primes: 2, 3, 5, 7, 11, 13, 17, 19, 23, 29, 31, 37, 41, 43, 47, 53, 59, 61, 67.

FUEL FOR THOUGHT

NOTICE THAT THERE are some pairs of primes that differ by only 2, like 11 and 13 or 59 and 61. These pairs of primes are called **twin primes**.

HOW DO YOU FACTOR WHOLE NUMBERS?

Every whole number (except 0 and 1) is either prime or the product of primes. For example, 11 is a prime and 12 is the product of primes ($2 \times 2 \times 3 = 12$).

INSIDE TRACK

DIVISIBILITY RULES FOR 2, 3, 5, AND 7

Divisibility rule for 2: Look only at the units digit. If the units digit of the number is 0, 2, 4, 6, or 8, then the number is even, i.e., the number is divisible by 2. The number 35,976 is even because its unit digit is 6.

Divisibility rule for 3: Look at all the digits. If the sum of all the digits is divisible by 3, then the number is divisible by 3. For instance, 63,156 is divisible by 3 because the sum of its digits, which is 21, is divisible by 3.

Divisibility rule for 5: Look only at the units digit. If the units digit of the number is 0 or 5, then the number is divisible by 5. For example, 745,980 is divisible by 5 because its units digit is 0.

Divisibility rule for 7: Look at the whole number. If the number is less than 70, see if you recognize the number as a multiple of 7 (7, 14, 21, 28, 35, 42, 49, 56, 63, 70). For a two-digit number like 91, subtract out 70 and see whether the result ($91 - 70 = 21$) is a multiple of 7; 91 is divisible by 7 because when you subtract 70 from 91, you get 21, which is a multiple of 7.

How can you factor any whole number less than 25? Test for divisibility by 2 and by 3. If the number that is less than 25 is not divisible by 2 and it is not divisible by 3, then it is prime.

Example

Factor 12 into its prime factors. Is 12 divisible by 2? Yes, $12 \div 2 = 6$. Write down the factor 2. Is 6 divisible by 2? Yes, $6 \div 2 = 3$. Write down the factor 2 again; 3 is prime. You now know that $12 = 2 \times 2 \times 3$.

FUEL FOR THOUGHT

WHEN YOU WRITE $3 \times 5 = 15$, the 3 and the 5 are **factors** and the 15 is the **product**.

The word *factor* works as a verb as well as a noun. When you figure out that the *factors* of 15 are 5 and 3, you are *factoring.*

Example

Factor 13 into its prime factors. Is 13 divisible by 2? No. Is 13 divisible by 3? No. Then, remember, it's prime.

FUEL FOR THOUGHT

The number 15 is **divisible** by 3 because 15 has no remainder when divided by 3; 3 goes into 15 evenly.

A number like 12 that has 2 as a factor is **divisible** by 2. Such a number is an **even** number. A number like 13 that leaves a remainder when divided by 2 is an **odd** number.

How can you factor any whole number less than 49? Test for divisibility by 2, 3, and 5. If the number is not divisible by 2, 3, or 5, it is prime.

Examples

Factor 21 into its prime factors. Is 21 divisible by 2? No. Is 21 divisible by 3? Yes. Write down the 3: $21 \div 3 = 7$; 7 is prime, so the factors of 21 are 3 and 7.

Factor 43 into its prime factors. Is 43 divisible by 2? No. Is 43 divisible by 3? No. Is 43 divisible by 5? No, so 43 is a prime number.

How can you factor any whole number less than 121? Test for divisibility by 2, 3, 5, and 7. If the number is not divisible by 2, 3, 5, or 7, then it is prime.

Example

Factor 89 into its prime factors. Is 89 divisible by 2? No. Is 89 divisible by 3? Again, no. Is 89 divisible by 5? No; almost done. Is 89 divisible by 7? No—okay, so 89 is a prime number.

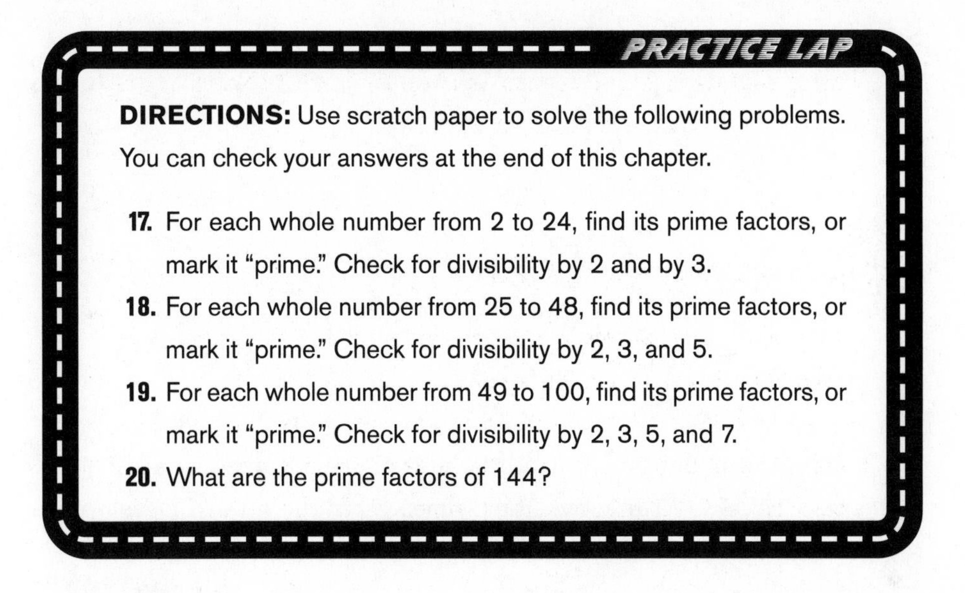

PRACTICE LAP

DIRECTIONS: Use scratch paper to solve the following problems. You can check your answers at the end of this chapter.

17. For each whole number from 2 to 24, find its prime factors, or mark it "prime." Check for divisibility by 2 and by 3.

18. For each whole number from 25 to 48, find its prime factors, or mark it "prime." Check for divisibility by 2, 3, and 5.

19. For each whole number from 49 to 100, find its prime factors, or mark it "prime." Check for divisibility by 2, 3, 5, and 7.

20. What are the prime factors of 144?

ANSWERS

1. 1,234,567,890

2. 9 billion, 876 million, 543 thousand, 210.

3. The value of the 7 in the number 9,876,543,210 is 70,000,000 (seventy million).

4. 8 + 7 = 15; perhaps you flashed on this. If you have not yet achieved flash on this addition fact, here are three ways of figuring it out. Let's say that this problem represents a pile of 8 pennies being combined with a pile of 7 pennies.

Method #1: Slide two of the pennies from the 7 pile over to the 8 pile. You now have a pile of 5 and a pile of 10: 10 + 5 = 15.

Method #2: The pile of 8 pennies is worth a nickel and 3 pennies, and the pile of 7 pennies is worth a nickel and 2 pennies. The pennies make 3 + 2 = 5, and the 2 nickels make 5 + 5 = 10; 10 + 5 = 15.

Method #3: 8 + 7 is 1 more than 7 + 7. Flash on 7 + 7 = 14 (doubles are easy to learn for many people), and one more than 14 is 15.

5. 347 + 892 = 1,239

Method #1: Write the numbers down under each other.

$$\begin{array}{r} {}^{1}\\ 347 \\ +\ 892 \\ \hline 1{,}239 \end{array}$$

6.

$$\begin{array}{r} \overset{3\,4}{935} \\ 219 \\ 47 \\ 272 \\ 518 \\ 794 \\ 333 \\ +\ \ 25 \\ \hline 3{,}143 \end{array}$$

For this problem, you can add the unit digits as you go: 5, 14, 21, 23, 31, 35, 38, 43. Write down the 3 ones, carry the 4 tens. Developing answer: 3.

Add the tens digits (including the carry of 4): 4, 7, 8, 12, 19, 20, 29, 32, 34. Write down the 4 tens, carry the 3 hundreds. Developing answer: 43.

Add the hundreds digits (starting with the carry): 3, 12, 14, 14, 16, 21, 28, 31, 31. Write down the 31. The answer is 3,143.

7. $12 - 7 = 5$

Did you flash on this? If not, do you flash on $7 + 5 = 12$? A way to figure out $7 + 5 = 12$ is to think of the 7 as 5 plus 2, so $7 + 5$ is $(5 + 2) + 5 = 5 + 5 + 2$, which is $10 + 2 = 12$.

8.

$$\begin{array}{r} \overset{7\ 13\ 10}{\not{8}\,\not{4}\,\not{0}} \\ -\,1\,6\,2 \\ \hline 6\,7\,8 \end{array}$$

9. The children who get off the bus at the first three schools are $12 + 7 + 15 = 34$. Remaining on the bus are $43 - 34 = 9$ children.

10. $5 \times 8 = 40$

11. $9 \times 7 = 63$

12. $348 \times 624 = 217{,}152$

$$\begin{array}{r} 348 \\ \times\ 624 \\ \hline 1392 \\ 696 \\ 2088 \\ \hline 217{,}152 \end{array}$$

13. $17 \times 7 = 119$. Mary has 119 marbles.

14. $56 \div 7 = 8$

15. How many times can you subtract 3 from 681,087? This is a division problem. The answer is achieved by dividing 3 into 681,087. The answer is 227,029. The number of times you can subtract 3 from 681,087 is 227,029.

16. $150{,}000 \div 52 = 2{,}884$ R32. To verify this answer, think of $(52 \times 2{,}884) + 32$.

17. 2: prime
3: prime
4: $2 \times 2 = 2^2$
5: prime
6: 2×3
7: prime
8: $2 \times 2 \times 2 = 2^3$
9: $3 \times 3 = 3^2$
10: 2×5
11: prime
12: $2 \times 2 \times 3 = 2^2 \times 3$
13: prime
14: 2×7
15: 3×5
16: $2 \times 2 \times 2 \times 2 = 2^4$
17: prime
18: $2 \times 3 \times 3 = 2 \times 3^2$
19: prime
20: $2 \times 2 \times 5 = 2^2 \times 5$
21: 3×7
22: 2×11
23: prime
24: $2 \times 2 \times 2 \times 3 = 2^3 \times 3$

18. 25: $5 \times 5 = 5^2$
26: 2×13
27: $3 \times 3 \times 3 = 3^3$
28: $2 \times 2 \times 7 = 2^2 \times 7$
29: prime
30: $2 \times 3 \times 5$
31: prime
32: $2 \times 2 \times 2 \times 2 \times 2 = 2^5$
33: 3×11
34: 2×17
35: 5×7
36: $2 \times 2 \times 3 \times 3 = 2^2 \times 3^2$
37: prime
38: 2×19
39: 3×13
40: $2 \times 2 \times 2 \times 5 = 2^3 \times 5$
41: prime
42: $2 \times 3 \times 7$
43: prime
44: $2 \times 2 \times 11 = 2^2 \times 11$
45: $3 \times 3 \times 5 = 3^2 \times 5$
46: 2×23
47: prime
48: $2 \times 2 \times 2 \times 2 \times 3 = 2^4 \times 3$

19. 49: $7 \times 7 = 7^2$
50: $2 \times 5 \times 5 = 2 \times 5^2$
51: 3×17
52: $2 \times 2 \times 13 = 2^2 \times 13$
53: prime
54: $2 \times 3 \times 3 \times 3 = 2 \times 3^3$
55: 5×11
56: $2 \times 2 \times 2 \times 7 = 2^3 \times 7$
57: 3×19
58: 2×29
59: prime
60: $2 \times 2 \times 3 \times 5 = 2^2 \times 3 \times 5$
61: prime
62: 2×31
63: $3 \times 3 \times 7 = 3^2 \times 7$
64: $2 \times 2 \times 2 \times 2 \times 2 \times 2 = 2^6$
65: 5×13
66: $2 \times 3 \times 11$

67: prime
68: $2 \times 2 \times 17 = 2^2 \times 17$
69: 3×23
70: $2 \times 5 \times 7$
71: prime
72: $2 \times 2 \times 2 \times 3 \times 3$
$= 2^3 \times 3^2$
73: prime
74: 2×37
75: $3 \times 5 \times 5 = 3 \times 5^2$
76: $2 \times 2 \times 19 = 2^2 \times 19$
77: 7×11
78: $2 \times 3 \times 13$
79: prime
80: $2 \times 2 \times 2 \times 2 \times 5 = 2^4 \times 5$
81: $3 \times 3 \times 3 \times 3 = 3^4$
82: 2×41
83: prime
84: $2 \times 2 \times 3 \times 7 = 2^2 \times 3 \times 7$
85: 5×17
86: 2×43
87: 3×29
88: $2 \times 2 \times 2 \times 11 = 2^3 \times 11$
89: prime
90: $2 \times 3 \times 3 \times 5 = 2 \times 3^2 \times 5$
91: 7×13
92: $2 \times 2 \times 23 = 2^2 \times 23$
93: 3×31
94: 2×47
95: 5×19
96: $2 \times 2 \times 2 \times 2 \times 2 \times 3$
$= 2^5 \times 3$
97: prime
98: $2 \times 7 \times 7 = 2 \times 7^2$
99: $3 \times 3 \times 11 = 3^2 \times 11$
100: $2 \times 2 \times 5 \times 5 = 2^2 \times 5^2$

20. What are the prime factors of 144?
Is 144 divisible by 2? Yes. Developing answer: 2, so $144 \div 2 = 72$.
Is 72 divisible by 2? Yes. Developing answer: 2×2, so $72 \div 2 = 36$.
Is 36 divisible by 2? Yes. Developing answer: $2 \times 2 \times 2$, so $36 \div 2 = 18$.
Is 18 divisible by 2? Yes. Developing answer: $2 \times 2 \times 2 \times 2$, so $18 \div 2 = 9$.
Is 9 divisible by 2? No. Is it divisible by 3? Yes. Developing answer: $2 \times 2 \times 2 \times 2 \times 3$, so $9 \div 3 = 3$; 3 is prime.
Final answer: $2 \times 2 \times 2 \times 2 \times 3 \times 3 = 2^4 \times 3^2$.

4 Fractions and Mixed Numbers

WHAT'S AROUND THE BEND

- What's a Fraction?
- Changing a Fraction without Changing Its Value
- Addition and Subtraction with Fractions
- Multiplying and Dividing Fractions
- Reciprocals
- Reciprocals on the Number Line
- Introducing Mixed Numbers
- Converting to a Fraction
- Converting a Fraction to a Mixed Number
- Adding and Subtracting Mixed Numbers
- Multiplying and Dividing Mixed Numbers

Fractions are used to represent parts of a whole. You can think of the fraction bar as meaning "out of." You can also think of the fraction bar as meaning "divided by."

Although most people call the top part of the fraction the "top" and the bottom part of a fraction the "bottom," the technical names are **numerator** and **denominator**.

CAUTION!

YOU ARE NEVER allowed to have a zero in the denominator. A fraction whose denominator is zero is undefined.

A proper fraction has a numerator that is smaller than its denominator. Examples are $\frac{1}{2}$, $\frac{13}{25}$, and $\frac{99}{100}$. Improper fractions have numerators that are bigger than their denominators. Examples include $\frac{8}{5}$, $\frac{99}{13}$, and $\frac{5}{2}$.

What if the numerator and denominator are equal (making the fraction equal to 1), as is the case with $\frac{2}{2}$, $\frac{9}{9}$, and $\frac{20}{20}$? Are these proper or improper fractions? The rule is that these fractions must be called **improper fractions**.

A fraction that represents a particular part of the whole is sometimes referred to as a **fractional part**. For example, let's say that a family has 4 cats and 2 dogs. What fractional part of their pets are cats? Because 4 out of the total 6 animals are cats, the fractional part of their pets that are cats is equal to $\frac{4}{6}$. You can reduce this fraction to $\frac{2}{3}$: $\frac{4 \div 2}{6 \div 2} = \frac{2}{3}$.

FUEL FOR THOUGHT

A **proper fraction** (with positive numerator and denominator) has a value that is less than 1. If the value is 1 or greater, the fraction is called an **improper fraction**.

You can identify an **improper fraction** easily: Its numerator is as big as or bigger than its denominator. Is that a no-no? No, it is not a no-no. There is really *nothing* improper about an improper fraction.

There's one more term you need to know, and then we can stop talking about fractions and start using them. That term is **mixed number**, which consists of a whole number paired with a proper fraction. Examples would include $3\frac{3}{4}$, $1\frac{5}{8}$, and $4\frac{2}{3}$.

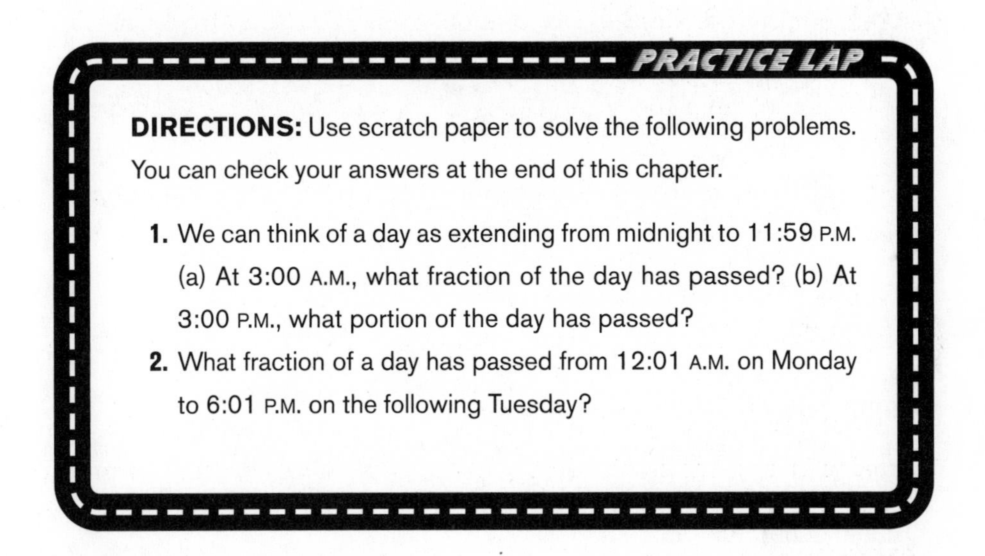

WORKING WITH FRACTIONS

What is it that doesn't get smaller when you reduce it? A fraction! There are two operations you can do to change the look and feel of a fraction without changing its value: You can reduce it (as in converting $\frac{6}{8}$ to $\frac{3}{4}$) or you can augment it (as in converting $\frac{3}{4}$ to $\frac{6}{8}$).

Reducing a Fraction

When you reduce $\frac{6}{8}$ to $\frac{3}{4}$, you don't change the value of the fraction. How is it possible to change the fraction without changing its value?

Well, both $\frac{6}{8}$ and $\frac{3}{4}$ reside at the same point on the number line. That means $\frac{6}{8}$ is the same number as $\frac{3}{4}$. A piece of lumber that is $\frac{6}{8}$ of an inch thick has the same thickness as a piece of lumber that is $\frac{3}{4}$ of an inch thick. The principle used to preserve the value of the fraction was to divide the numerator and the denominator by the same number (in this case, 2).

When you reduce a fraction, you preserve its value.

Example

Reduce $\frac{58}{87}$.

Let's see: 58 is divisible by 2, but not by 3, and 87 is divisible by 3, but not by 2. What can you do?

As you learned in Chapter 3, you can express whole numbers as the product of their prime factors. So 58 works out to 2×29 and 87 works out to 3×29. So your fraction is $\frac{58}{87} = \frac{(2 \times 29)}{(3 \times 29)}$. Now you can divide both the numerator and the denominator by 29. (This is called "canceling" the 29.) The result is $\frac{58}{87} = \frac{(2 \times 29)}{(3 \times 29)} = \frac{2}{3}$.

For any fraction you might encounter, expressed as a whole number divided by a whole number, you can write both whole numbers as the product of their prime factors, like with $\frac{58}{87}$. Then, cancel, cancel, cancel until there are no more primes to cancel.

Augmenting a Fraction

Sometimes you need to do the opposite of reducing. Instead of converting $\frac{6}{8}$ to $\frac{3}{4}$, you may need to convert $\frac{3}{4}$ to $\frac{6}{8}$. You do this by multiplying the numerator and denominator by 2.

You can augment $\frac{3}{4}$ of a pizza by using a pizza cutter. Suppose the full pizza has been sliced into four equal slices, and three of the slices ($\frac{3}{4}$ of the pizza) are topped with mushrooms. Cut each of the four slices in half. There are now 8 equal slices, so each slice is $\frac{1}{8}$ of the pizza. Six of those slices are topped with mushrooms. (The three slices of mushroom pizza became six slices when you did the cutting.) That means that $\frac{6}{8}$ of the pizza is topped with mushrooms—the same $\frac{3}{4}$ of the pizza that was topped with mushrooms before you used the pizza cutter.

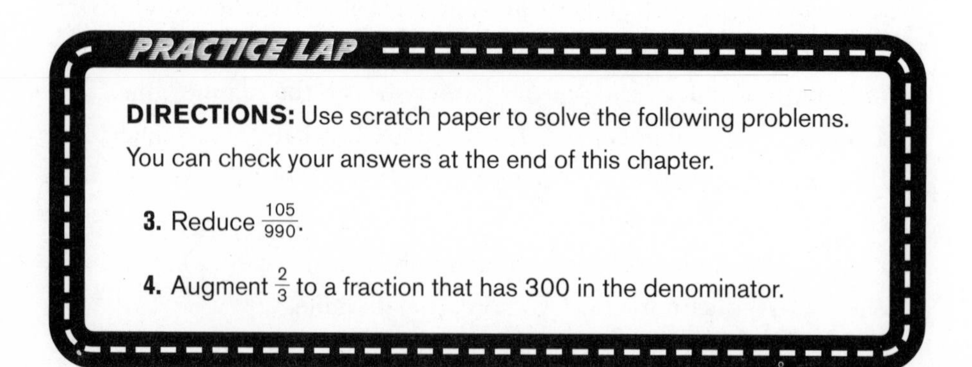

PRACTICE LAP

DIRECTIONS: Use scratch paper to solve the following problems. You can check your answers at the end of this chapter.

3. Reduce $\frac{105}{990}$.

4. Augment $\frac{2}{3}$ to a fraction that has 300 in the denominator.

Adding Fractions

Do you have any loose change? A friend would like to borrow a quarter. Do you happen to have another quarter he can borrow? Don't worry; it's just a loan. And while you're at it, let him borrow still another quarter. All right, how many quarters does he owe you?

If he borrowed one quarter from you, then another quarter, and then still another quarter, he borrowed three quarters from you. In other words, he borrowed $\frac{1}{4} + \frac{1}{4} + \frac{1}{4}$, or a total of $\frac{3}{4}$.

Here's another question: How much is $\frac{1}{10} + \frac{1}{10} + \frac{1}{10}$? It's $\frac{3}{10}$. And how much is $\frac{2}{9} + \frac{1}{9} + \frac{2}{9} + \frac{2}{9}$? Go ahead and add them up. It's $\frac{7}{9}$. When you add fractions with the same denominator, all you have to do is add the numerators.

Now, how much is $\frac{1}{6} + \frac{1}{6} + \frac{1}{6}$? It's $\frac{3}{6}$. But you can reduce that to $\frac{1}{2}$. What did you really do just then? You divided the numerator (3) by 3 and you divided the denominator (6) by 3. There's a law of arithmetic that says when you divide the top of a fraction by any number, you must also divide the bottom of that fraction by the same number.

Now add together $\frac{1}{2} + \frac{1}{2} + \frac{1}{2} + \frac{1}{2}$. What did you come up with? Was it 2? You did this: $\frac{1}{2} + \frac{1}{2} + \frac{1}{2} + \frac{1}{2} = \frac{4}{2} = 2$.

In order to add fractions, they must have the same denominator:

$$\frac{3}{11} + \frac{4}{11} = \frac{\text{numerator + numerator}}{\text{denominator}} = \frac{7}{11}$$

INSIDE TRACK

SHOULD YOU REDUCE your fractions to lowest possible terms? If you left $\frac{5}{5}$ as is (instead of making it 1), is it wrong? No, but by convention, we always reduce our fractions as much as possible. Indeed, there are mathematicians who can't go to sleep at night unless they're sure that every fraction has been reduced to its lowest possible terms. You probably wouldn't want to keep these poor people up all night, so always reduce your fractions.

But with Unlike Denominators . . .

Have you ever heard the expression, "That's like adding apples to oranges?" You can add apples to apples; you can add oranges to oranges. But you can't add apples *to* oranges.

Can you add $\frac{1}{2}$ and $\frac{1}{3}$? Believe it or not, you can. The problem is that they don't have the same common denominator. You need to give them a common denominator before they can be added. Do you have any idea how to do this?

You need to convert $\frac{1}{2}$ into $\frac{3}{6}$ and $\frac{1}{3}$ into $\frac{2}{6}$. Here's how to do it:

$$\frac{1 \times 3}{2 \times 3} + \frac{1 \times 2}{3 \times 2} = \frac{3}{6} + \frac{2}{6} = \frac{5}{6}$$

Once the fractions have a common denominator, you can add them: $\frac{3}{6} + \frac{2}{6} = \frac{5}{6}$.

CAUTION!

REMEMBER THE OLD arithmetic law: What you do to the bottom of a fraction (the denominator) you must also do to the top (the numerator).

Example

$$\frac{1}{4} + \frac{1}{6} + \frac{5}{8} =$$

What is the smallest denominator you can use? You need to augment the three fractions so that they all have the same denominator. In other words, you need to find the LCD (least common denominator) for 4, 6, and 8.

One way to find the LCD is to run the multiples of 4, 6, and 8, looking for the first number that appears in all three lists:

Multiples of 4: 4, 8, 12, 16, 20, 24, . . .
Multiples of 6: 6, 12, 18, 24, . . .
Multiples of 8: 8, 16, 24, . . .

The first number that appears on all three lists is 24, so 24 is your least common denominator.

Another way of finding the LCD is to write each of the denominators, 4, 6, and 8, as the product of its prime factors:

$$4 = 2 \times 2$$
$$6 = 2 \times 3$$
$$8 = 2 \times 2 \times 2$$

The number you want needs to have three 2's (to accommodate the 8) and one 3 (to accommodate the 6). So, the LCD is $2 \times 2 \times 2 \times 3$, which is 24.

You can now augment each of the fractions to a fraction in which the denominator is 24:

$$\frac{1}{4} = \frac{6}{24}$$
$$\frac{1}{6} = \frac{4}{24}$$
$$\frac{5}{8} = \frac{15}{24}$$
$$\frac{1}{4} + \frac{1}{6} + \frac{5}{8} = \frac{6}{24} + \frac{4}{24} + \frac{15}{24} = \frac{25}{24}\text{, or } 1\frac{1}{24}$$

PRACTICE LAP

DIRECTIONS: Use scratch paper to solve the following problems. You can check your answers at the end of this chapter.

5. $\frac{3}{8} + \frac{5}{8} =$

6. $\frac{3}{8} + \frac{4}{5} + \frac{7}{3} + \frac{9}{10} =$

How to Subtract Fractions

Subtracting fractions is not much different from adding them, except for a change of sign. If the denominators are the same, we subtract the numerators.

Example

$\frac{5}{8} - \frac{2}{8} =$

The denominators are the same, so do the subtraction: $5 - 2 = 3$. The answer is $\frac{3}{8}$.

If the denominators are different, you augment the fractions, just like when you add fractions, so that they both have the same denominator. Then, subtract the numerators.

Example

$\frac{5}{8} - \frac{1}{4} =$

Augment the $\frac{1}{4}$: $\frac{1}{4} = \frac{2}{8}$. Your problem is now $\frac{5}{8} - \frac{1}{4} = \frac{5}{8} - \frac{2}{8} = \frac{3}{8}$.

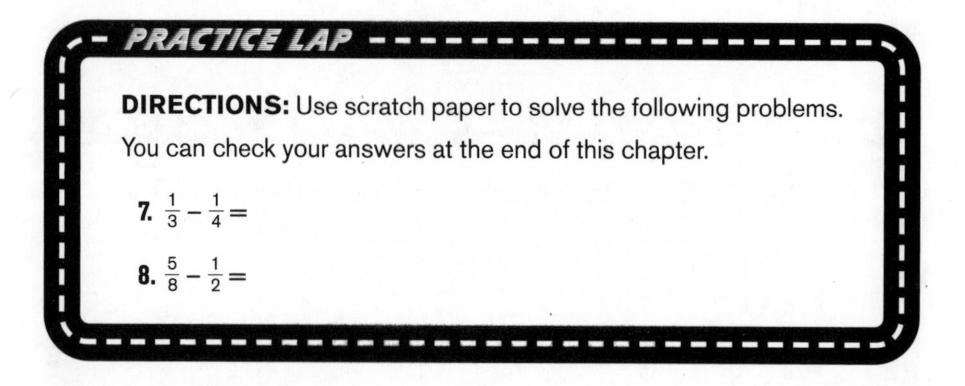

PRACTICE LAP

DIRECTIONS: Use scratch paper to solve the following problems. You can check your answers at the end of this chapter.

7. $\frac{1}{3} - \frac{1}{4} =$

8. $\frac{5}{8} - \frac{1}{2} =$

How to Multiply Fractions

You'll find that multiplying fractions is different from adding and subtracting them because you don't need to find a common denominator before you do the math operation. Actually, this makes multiplying fractions easier than adding or subtracting them.

Example

$\frac{3}{5} \times \frac{5}{8}$

Multiply both the numerators and denominators straight across.

$\frac{3}{5} \times \frac{5}{8} = \frac{(3 \times 5)}{(5 \times 8)} = \frac{15}{40}$, which can be reduced to $\frac{3}{8}$.

FUEL FOR THOUGHT

PROBLEM: How much is one-third of one-eighth?

Solution: $\frac{1}{3} \times \frac{1}{8} = \frac{1}{24}$

You can see by the way this problem is worded that *of* means "multiply," or "times." The question would be the same if it were, "How much is one-third times one-eighth?"

Canceling Out

When you multiply fractions, you can often save time and mental energy by cancelling out. Here's how it works.

Example

How much is $\frac{5}{6} \times \frac{3}{4}$?

$\frac{5}{6} \times \frac{3}{4} = \frac{5}{\not{6}_{2}} \times \frac{\not{3}^{1}}{4} = \frac{5}{8}$

You divided the 6 in $\frac{5}{6}$ by 3 and you divided the 3 in $\frac{3}{4}$ by 3. In other words, the 3 in the 6 and the 3 in the 3 cancelled each other out.

Cancelling out helps you reduce fractions to their lowest possible terms. While there's no law of arithmetic that says you have to do this, it makes multiplication easier because it's much easier to work with smaller numbers.

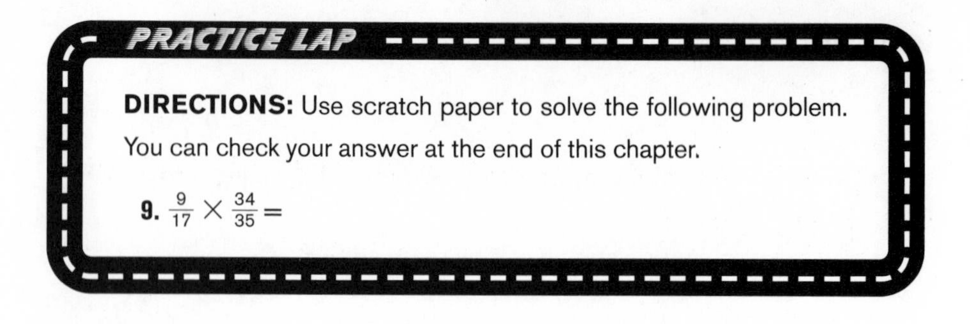

PRACTICE LAP

DIRECTIONS: Use scratch paper to solve the following problem. You can check your answer at the end of this chapter.

9. $\frac{9}{17} \times \frac{34}{35} =$

Dividing Fractions

The division of fractions is like multiplication, but with a twist. You'll find the trick is to turn a division problem into a multiplication problem.

Let's get right into it. When a fraction problem says to divide, just say no. When a fraction problem asks you to divide the first fraction by a second fraction, you say "no" to the division. Instead, you take the first fraction and *multiply* it by the reciprocal of the second fraction.

The **reciprocal** of a fraction is found by turning the fraction upside down. The reciprocal of 4 is $\frac{1}{4}$, and the reciprocal of $\frac{1}{4}$ is 4. In other words, 4 and $\frac{1}{4}$ are reciprocals of each other.

The reciprocal of -4 is $-\frac{1}{4}$ and the reciprocal of $-\frac{1}{4}$ is -4. In other words, -4 and $-\frac{1}{4}$ are reciprocals of each other.

Reciprocals come in pairs, and the numbers in a reciprocal pair are either both negative or both positive. This is because two negative reciprocals multiply to $+1$ and two positive reciprocals multiply to $+1$.

FUEL FOR THOUGHT

ZERO IS THE only number that doesn't have a reciprocal. This is because any number times zero is zero, so there is no number to multiply zero by that will give you 1.

Example

$\frac{3}{5} \div \frac{5}{8} =$

The problem says, "Divide." Say, "No." Instead of dividing the first fraction by $\frac{5}{8}$, multiply it by $\frac{8}{5}$:

$\frac{3}{5} \times \frac{8}{5} = \frac{(3 \times 8)}{(5 \times 5)} = \frac{24}{25}$

CAUTION!

THE ORDER OF THE NUMBERS

WHEN YOU MULTIPLY two numbers, you get the same answer regardless of their order. When you divide one number by another, does it matter in which order the numbers appear? Let's find out.

What is $\frac{1}{3} \div \frac{1}{4}$?

$\frac{1}{3} \div \frac{1}{4} = \frac{1}{3} \times \frac{4}{1} = \frac{4}{3} = 1\frac{1}{3}$

Now, what is $\frac{1}{4} \div \frac{1}{3}$?

$\frac{1}{4} \div \frac{1}{3} = \frac{1}{4} \div \frac{3}{1} = \frac{3}{4}$

There is no way that $\frac{3}{4}$ can equal $1\frac{1}{3}$. So, when you do division of fractions, you must be very careful about the order of the numbers. The number that is being divided always comes before the division sign, and the number doing the dividing always comes after the division sign.

PRACTICE LAP

DIRECTIONS: Use scratch paper to solve the following problems. You can check your answers at the end of this chapter.

10. $\frac{3}{5} \div \frac{3}{8} =$

11. How much is $\frac{5}{6}$ divided by $\frac{2}{9}$?

12. $\frac{1}{4} \div \frac{5}{8} =$

13. What happens to the number 17 if you multiply it by 3 and then multiply that result by the reciprocal of 3?

14. What happens to the number 17 if you multiply it by any positive number p, and then multiply that result by the reciprocal of p?

INSIDE TRACK

ANY NUMBER DIVIDED by itself equals 1: $\frac{5}{8} \div \frac{5}{8} = 1$.

When you divide a number by a smaller number, the answer (quotient) will be greater than 1: $\frac{3}{4} \div \frac{2}{3} = 1\frac{1}{8}$.

When you divide a number by a larger number, the quotient will be less than 1: $\frac{2}{7} \div \frac{4}{5} = \frac{5}{14}$.

Which Is Larger, p or $\frac{1}{p}$?

Suppose p is a positive number, $p > 0$. Which is larger, p or $\frac{1}{p}$? Think about it for a moment.

Many people answer incorrectly that p is greater. They are thinking about a whole number, like 3. They know that 3 is greater than $\frac{1}{3}$. But you, dear reader, know better. You know that 3 is indeed greater than $\frac{1}{3}$, but you also know that p could be $\frac{1}{3}$. Then, the reciprocal of $\frac{1}{3}$, which is 3, is bigger.

For positive p, the question of which is bigger, p or $\frac{1}{p}$, could go either way.

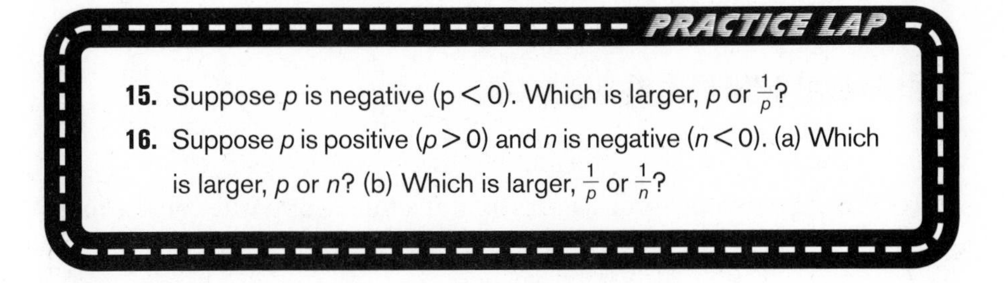

PRACTICE LAP

15. Suppose p is negative ($p < 0$). Which is larger, p or $\frac{1}{p}$?

16. Suppose p is positive ($p > 0$) and n is negative ($n < 0$). (a) Which is larger, p or n? (b) Which is larger, $\frac{1}{p}$ or $\frac{1}{n}$?

WHAT'S THE MIX-UP WITH MIXED NUMBERS?

An example of a mixed number is $5\frac{3}{4}$. This mixed number is a mixture, so to speak, of a whole number (5) and a fraction ($\frac{3}{4}$). The value of this mixed number is the sum, $5 + \frac{3}{4}$. When you write $5\frac{3}{4}$, you don't have to write the plus sign. It is implied.

Converting a Mixed Number to a Fraction

To convert mixed numbers into improper fractions on the fly, you just multiply the whole number by the denominator, add this to the numerator, and stick this value over the same denominator.

Look at the mixed number $5\frac{3}{4}$. To calculate the numerator of the new fraction, multiply the 5 by the 4 (20) and add 3 (23). The denominator of the new fraction is the 4. So, $5\frac{3}{4} = \frac{23}{4}$.

Now you know how to convert a mixed number to a fraction. That's pretty much all there is to it.

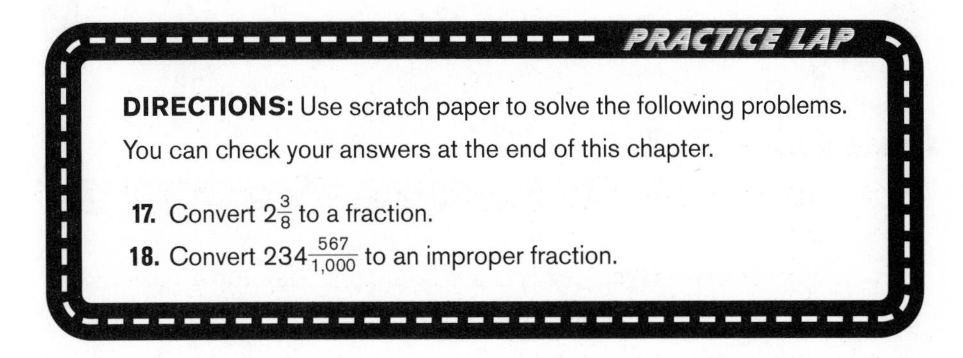

PRACTICE LAP

DIRECTIONS: Use scratch paper to solve the following problems. You can check your answers at the end of this chapter.

17. Convert $2\frac{3}{8}$ to a fraction.

18. Convert $234\frac{567}{1,000}$ to an improper fraction.

RECIPROCALS ON THE NUMBER LINE

How do reciprocals look on the number line?

For contrast with reciprocals, let's look at additive opposites on the number line:

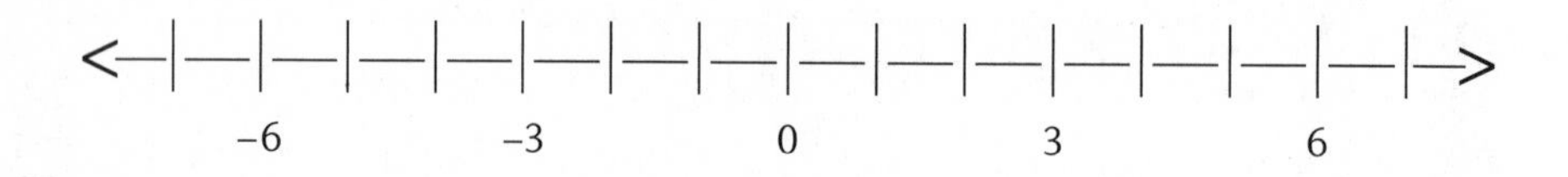

In the diagram, you see two pairs of opposites, 3 and –3, and 6 and –6. As you can see, the first pair, 3 and –3, reflect across the zero (as if there were a mirror at the zero), and the second pair, 6 and –6, also reflect across the zero.

For multiplicative opposites (reciprocals), the picture is quite different. Pairs of reciprocals are always both on the same side of the number line, but the *whole pattern* of positive pairs of reciprocals reflects across the zero. The relationship of negative reciprocals is the mirror image (across the zero) of the positive reciprocals.

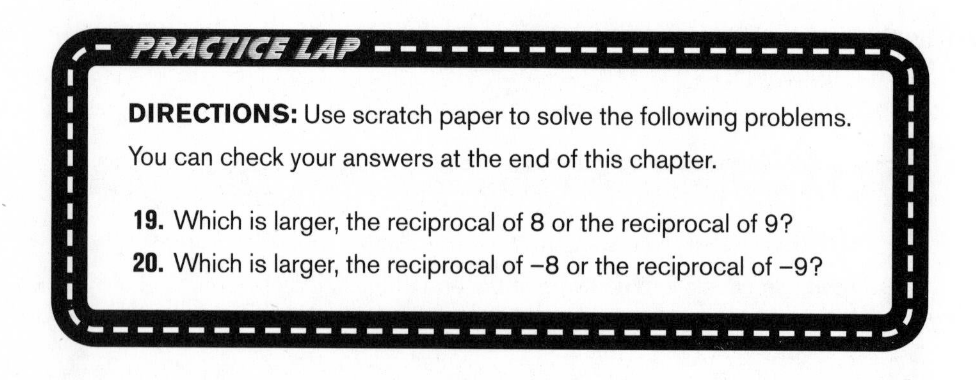

DIRECTIONS: Use scratch paper to solve the following problems. You can check your answers at the end of this chapter.

19. Which is larger, the reciprocal of 8 or the reciprocal of 9?

20. Which is larger, the reciprocal of –8 or the reciprocal of –9?

Converting a Fraction to a Mixed Number

To convert $\frac{23}{4}$ to a mixed number, divide the numerator by the denominator: $23 \div 4 = 5$ with remainder 3. The remainder represents the number of quarters, the new numerator of your fraction. The denominator remains 4. So, $\frac{23}{4} = 5\frac{3}{4}$.

Of course, if the division produces no remainder, you don't get a mixed number. The fraction $\frac{12}{3}$ converts to 4, which is a whole number, not a mixed number.

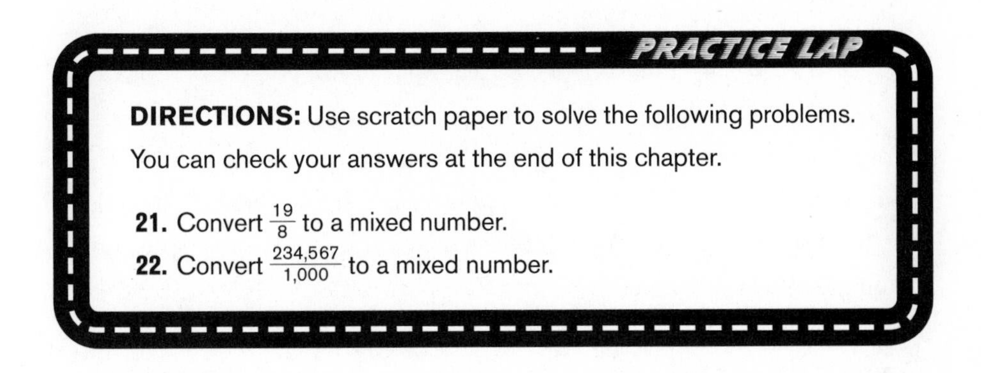

Adding Mixed Numbers

When you add two mixed numbers, you have two whole numbers and two fractions. Add the two whole numbers to get a whole number and add the two fractions to get a fraction.

Example

$$5\tfrac{1}{4} + 2\tfrac{3}{8} =$$

Add the whole numbers $(5 + 2 = 7)$ and add the fractions $(\frac{1}{4} + \frac{3}{8} = \frac{2}{8} + \frac{3}{8} = \frac{5}{8})$. The result is $5\frac{1}{4} + 2\frac{3}{8} = 7\frac{5}{8}$.

But, what if the fraction you get is an improper fraction?

Example

$$5\tfrac{3}{4} + 2\tfrac{7}{8} =$$

$$5\tfrac{3}{4} + 2\tfrac{7}{8} = 5 + \tfrac{6}{8} + 2 + \tfrac{7}{8} = 7 + \tfrac{13}{8} = 7\tfrac{13}{8}$$

This result satisfies the requirements of a mixed number, but you would be more satisfied if the fractional part, $\frac{13}{8}$, were a proper fraction. So, let's convert $\frac{13}{8}$ to a mixed number: $\frac{13}{8} = 1\frac{5}{8}$. Now you can write $7\frac{13}{8} = 7 + \frac{13}{8} = 7 + \frac{8}{8} + \frac{5}{8} = 7 + 1 + \frac{5}{8}$.

Add the 7 and the 1. Your result is $8\frac{5}{8}$, which is a mixed number with a proper fraction for its fractional part.

Example

$$5\frac{1}{4} + 2\frac{3}{4} =$$

$$5\frac{1}{4} + 2\frac{3}{4} = 5 + \frac{1}{4} + 2 + \frac{3}{4} = 7 + \frac{4}{4} = 7 + 1 = 8$$

You didn't get a mixed number, but that's all right. It is always true that a whole number plus a whole number equals a whole number, but there is no law that says that a mixed number plus a mixed number has to equal a mixed number.

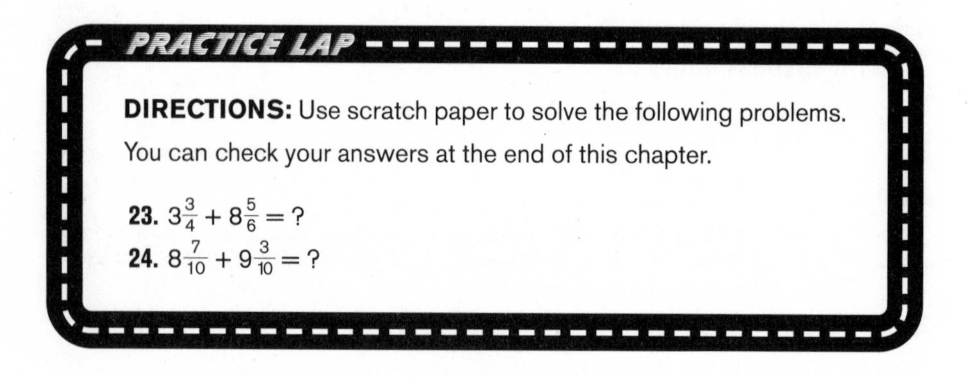

PRACTICE LAP

DIRECTIONS: Use scratch paper to solve the following problems. You can check your answers at the end of this chapter.

23. $3\frac{3}{4} + 8\frac{5}{6} = ?$

24. $8\frac{7}{10} + 9\frac{3}{10} = ?$

Subtracting Mixed Numbers

When you subtract two mixed numbers, you have two whole numbers and two mixed numbers, just like when you add two mixed numbers. So, you subtract the whole numbers and you subtract the mixed numbers.

Example

$$8\frac{5}{9} - 2\frac{1}{3} =$$

The whole number part of the result is $8 - 2 = 6$. The fractional part of the result is $\frac{5}{9} - \frac{1}{3} = \frac{5}{9} - \frac{3}{9} = \frac{2}{9}$. The result is $6\frac{2}{9}$.

It sometimes happens that the subtraction won't work with your fractions. That's easy to fix: Just slice up one of whole numbers.

Example

$$8\frac{3}{9} - 2\frac{5}{9} =$$

The subtraction of the fractions, $(\frac{3}{9} - \frac{5}{9})$, is not going to work. There are not enough ninths in $\frac{3}{9}$ to subtract $\frac{5}{9}$. So, slice up one of the 8 wholes into 9 ninths. So, $8\frac{3}{9} = 8 + \frac{3}{9}$ becomes $7 + 1 + \frac{3}{9} = 7 + \frac{9}{9} + \frac{3}{9} = 7 + \frac{12}{9} = 7\frac{12}{9}$.

Now, you have $7\frac{12}{9} - 2\frac{5}{9}$ and can subtract the whole numbers $(7 - 2 = 5)$ and subtract the fractions $(\frac{12}{9} - \frac{5}{9} = \frac{7}{9})$ to get the result of $5\frac{7}{9}$.

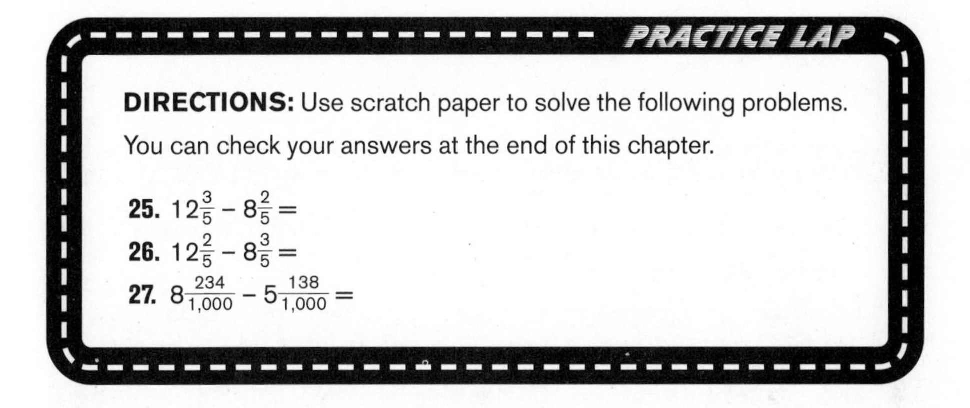

PRACTICE LAP

DIRECTIONS: Use scratch paper to solve the following problems. You can check your answers at the end of this chapter.

25. $12\frac{3}{5} - 8\frac{2}{5} =$

26. $12\frac{2}{5} - 8\frac{3}{5} =$

27. $8\frac{234}{1,000} - 5\frac{138}{1,000} =$

Multiplying and Dividing Mixed Numbers

You already know how to multiply and divide fractions. When you have the opportunity to do these operations on mixed numbers, the simplest thing is to convert the mixed numbers to fractions, and then multiply or divide.

Example

$$3\frac{2}{3} \times 2\frac{1}{4} =$$

$$3\frac{2}{3} \times 2\frac{1}{4} = \frac{11}{3} \times \frac{9}{4} = \frac{99}{12} = \frac{33}{4} = 8\frac{1}{4}$$

That was simple.

PRACTICE LAP

DIRECTIONS: Use scratch paper to solve the following problems. You can check your answers at the end of this chapter.

28. $5\frac{3}{4} \times 5\frac{3}{4} =$

29. $10\frac{4}{9} \div 5\frac{2}{9} =$

30. $8\frac{3}{4} \div 4 =$

ANSWERS

1. (a) $\frac{1}{8}$ of the day has passed. In other words, 3 hours of a 24-hour day have passed, and $\frac{3}{24}$ reduces to $\frac{1}{8}$. You can divide 24 hours into 8 equal portions of 3 hours each. (b) $\frac{5}{8}$ of the day has passed. From midnight to 3:00 P.M., five three-hour periods (that is to say, five-eighths of the day) have passed. The answer to the question is $\frac{5}{8}$.
2. The interval in question is 42 hours long. There are 24 hours in a day and $\frac{42}{24}$ reduces to $\frac{7}{4}$. The period in question is $\frac{7}{4}$ of a day, or $1\frac{3}{4}$ days.
3. Reduce the numerator. Is 105 divisible by 2? No. Is 105 divisible by 3? Yes. Developing answer: 3.
   $\frac{105}{3} = 35$. Is 35 divisible by 5? Yes. Developing answer: 3×5.
   $\frac{35}{5} = 7$. Is 7 prime? Yes.
   Reduced numerator: $105 = 3 \times 5 \times 7$.
   Reduce the denominator. Is 990 divisible by 2? Yes. Developing answer: 2.
   $\frac{990}{2} = 495$. Is 495 divisible by 2? No. Is 495 divisible by 3? Yes. Developing answer: 2×3; $\frac{495}{3} = 165$. Is 165 divisible by 3? Yes. Developing answer: $2 \times 3 \times 3$.
   $\frac{165}{3} = 55$. Is 55 divisible by 3? No. Is 55 divisible by 5? Yes. Developing answer: $2 \times 3 \times 3 \times 5$.
   $\frac{55}{5} = 11$. Is 11 prime? Yes.
   Reduced denominator: $990 = 2 \times 3 \times 3 \times 5 \times 11 = 2 \times 3^2 \times 5 \times 11$.

$\frac{105}{990} = \frac{(3 \times 5 \times 7)}{(2 \times 3 \times 3 \times 5 \times 11)}$. Cancel the 3 and the 5 from both the numerator and denominator. Final answer: $\frac{105}{990} = \frac{7}{(2 \times 3 \times 11)} = \frac{7}{66}$.

4. To get from 3 to 300, you multiply the denominator by 100. Do the same to the numerator. Multiply the numerator 2 by 100: $2 \times 100 = 200$. The answer is $\frac{200}{300}$.

5. $\frac{3}{8} + \frac{5}{8} = \frac{8}{8} = \frac{1}{1} = 1$

6. The denominators are 8, 5, 3, and 10. Find the LCM (least common multiple) of these numbers. We need three 2's to accommodate the 8, and we need a 3 and a 5. (We don't need an additional 2 and 5 for the number 10, because we already have a 2 and a 5.) So the LCM, which will serve as our LCD (least common denominator), is $2 \times 2 \times 2 \times 3 \times 5 = 120$. Augment $\frac{3}{8}$ to $\frac{45}{120}$, $\frac{4}{5}$ to $\frac{96}{120}$, $\frac{7}{3}$ to $\frac{280}{120}$, and $\frac{9}{10}$ to $\frac{108}{120}$. The result is $\frac{(45 + 96 + 280 + 108)}{120} = \frac{529}{120}$.

7. The least common denominator is 12: $\frac{1}{3} - \frac{1}{4} = \frac{4}{12} - \frac{3}{12} = \frac{1}{12}$

8. The least common denominator is 8: $\frac{5}{8} - \frac{1}{2} = \frac{5}{8} - \frac{4}{8} = \frac{1}{8}$

9. The hard way: Multiply the numerators and the denominators. $\frac{9}{17} \times \frac{34}{35} = \frac{(9 \times 34)}{(17 \times 35)} = \frac{306}{595}$, which reduces to $\frac{18}{35}$.

A simpler way is to factor the numerator and the denominator and cancel before multiplying. $\frac{(9 \times 34)}{(17 \times 35)} = \frac{(3 \times 3 \times 17 \times 2)}{(17 \times 5 \times 7)} = \frac{(3 \times 3 \times 2)}{(5 \times 7)} = \frac{18}{35}$.

10. $\frac{3}{5} \div \frac{3}{8} = \frac{\overset{1}{\cancel{3}}}{5} \times \frac{8}{\underset{1}{\cancel{3}}} = \frac{8}{5} = 1\frac{3}{5}$

11. $\frac{5}{6} \div \frac{2}{9} = \frac{5}{\underset{2}{\cancel{6}}} \times \frac{\overset{3}{\cancel{9}}}{2} = \frac{15}{4} = 3\frac{3}{4}$

12. $\frac{1}{4} \div \frac{5}{8} = \frac{1}{\underset{1}{\cancel{4}}} \times \frac{\overset{2}{\cancel{8}}}{5} = \frac{2}{5}$

13. $17 \cdot 3 = 51$; $51 \times (\frac{1}{3}) = 17$

14. $(17 \cdot p)(\frac{1}{p}) = 17[p(\frac{1}{p})] = 17(1) = 17$

15. If p is -3, its reciprocal, $-\frac{1}{3}$, is larger (more to the right on the number line). But if p is $-\frac{1}{3}$, its reciprocal, -3, is smaller (more to the left on the number line).

16. (a) p is larger than n ($p > n$), because any positive number is larger (more to the right on the number line) than any negative number. (b) $\frac{1}{p}$ is positive and $\frac{1}{n}$ is negative, so $\frac{1}{p}$ is larger than $\frac{1}{n}$ ($\frac{1}{p} > \frac{1}{n}$).

17. To find the numerator of your answer, multiply the whole number by the denominator of the fraction and add the numerator: $2 \times 8 + 3 = 19$. The denominator is 8. The result is $\frac{19}{8}$.

18. To find the numerator of your answer, multiply the whole number by the denominator of the fraction and add the numerator: $234 \times 1{,}000 + 567 = 234{,}000 + 567 = 234{,}567$. The denominator is 1,000. The result is $\frac{234{,}567}{1{,}000}$.

19. The question is, which is larger, $\frac{1}{8}$ or $\frac{1}{9}$? $\frac{1}{8} > \frac{1}{9}$.

20. The question is, which is larger, $-\frac{1}{8}$ or $-\frac{1}{9}$? $-\frac{1}{8} < -\frac{1}{9}$.

21. The division gives 2 with a remainder of 3; 3 is the numerator of your new fraction and the denominator remains 8. The answer is $2\frac{3}{8}$.

22. The division gives 234 with a remainder of 567. The answer is $234\frac{567}{1{,}000}$.

23. $3\frac{3}{4} + 8\frac{5}{6} = 3\frac{9}{12} + 8\frac{10}{12} = 11\frac{19}{12} = 11 + \frac{12}{12} + \frac{7}{12} = 11 + 1 + \frac{7}{12} = 12 + \frac{7}{12} = 12\frac{7}{12}$

24. $8\frac{7}{10} + 9\frac{3}{10} = 17 + \frac{10}{10} = 17 + 1 = 18$

25. $12\frac{3}{5} - 8\frac{2}{5} = 4\frac{1}{5}$

26. There are not enough fifths in $12\frac{2}{5}$, so slice up one of the 12 wholes: $12\frac{2}{5} = 12 + \frac{2}{5} = 11 + 1 + \frac{2}{5} = 11 + \frac{5}{5} + \frac{2}{5} = 11\frac{7}{5}$. Now, the subtraction problem can be written as $11\frac{7}{5} - 8\frac{3}{5} = (11 - 8) + (\frac{7}{5} - \frac{3}{5}) = 3 + \frac{4}{5} = 3\frac{4}{5}$.

27. $8\frac{234}{1{,}000} - 5\frac{138}{1{,}000} = 3\frac{96}{1{,}000} = 3\frac{12}{125}$

28. $5\frac{3}{4} \times 5\frac{3}{4} = \frac{23}{4} \times \frac{23}{4} = \frac{529}{16}$. This cannot be reduced because, with the numerator and denominator expressed as prime factors, $\frac{(23 \times 23)}{(2 \times 2 \times 2 \times 2)}$, there are no factors in common.

29. $\frac{94}{9} \div \frac{47}{9} = \frac{94}{\cancel{9}_1} \times \frac{\cancel{9}^1}{47} = \frac{94}{47} = \frac{(2 \times 47)}{47} = 2$

Now that you see that the answer is simply 2, look at the problem again and realize that $10\frac{4}{9}$ is indeed twice $5\frac{2}{9}$, because 10 is twice 5 and $\frac{4}{9}$ is twice $\frac{2}{9}$.

30. $8\frac{3}{4} \div 4 = \frac{35}{4} \div \frac{4}{1} = \frac{35}{4} \times \frac{1}{4} = \frac{35}{16} = 2\frac{3}{16}$. This result makes sense because the original problem is equivalent to $(8 + \frac{3}{4}) \div 4 = 2 + \frac{3}{16} = 2\frac{3}{16}$.

5 Dealing with Decimals

WHAT'S AROUND THE BEND

- What Is a Decimal?
- Padding a Decimal with Zeros
- Moving the Decimal Point to the Right and to the Left
- Arranging Decimals in Order
- Adding and Subtracting Decimals
- Rewriting a Decimal
- Multiplying Decimals
- Dividing Decimals

Decimals are really just fractions in disguise. Or maybe it's the fractions that are really decimals in disguise. You may never know who's disguised as what in this crazy world of mathematical espionage. One thing is for sure: You can equate fractions and decimals. For example, the fraction $\frac{1}{10}$ can be written as the decimal 0.1.

A decimal carries the decimal place-value system, which is based on the number 10, into the domain of fractions that are based on 10, namely $\frac{1}{10}$, $\frac{1}{100}$, $\frac{1}{1{,}000}$, etc.

To review the place-value system briefly, a whole number (e.g., 234) is the sum of its digits, each one multiplied, respectively, by 100, 10, and 1:

$$234 = (2 \times 100) + (3 \times 10) + (4 \times 1) = 200 + 30 + 4$$

A decimal expands on this concept, bringing in fractions as multipliers. The number 234.567 is equal to the sum of its digits, 2, 3, 4, 5, 6, and 7, each one multiplied, respectively, by 100, 10, 1, $\frac{1}{10}$, $\frac{1}{100}$, and $\frac{1}{1{,}000}$:

$$234.567 = (2 \times 100) + (3 \times 10) + (4 \times 1) + (5 \times \tfrac{1}{10}) + (6 \times \tfrac{1}{100}) + (7 \times \tfrac{1}{1{,}000}) = 200 + 30 + 4 + 0.5 + 0.06 + 0.007$$

In the following table, the places to the right of the decimal point are named for the number 0.987654321.

places to the right of the decimal point	digit from our example number 0.987654321	name of place	value of digit
1	9	tenths	0.9
2	8	hundredths	0.08
3	7	thousandths	0.007
4	6	ten thousandths	0.0006
5	5	hundred thousandths	0.00005
6	4	millionths	0.000004
7	3	ten millionths	0.0000003
8	2	hundred millionths	0.00000002
9	1	billionths	0.000000001

PRACTICE LAP

DIRECTIONS: Use scratch paper to solve the following problems. You can check your answers at the end of this chapter.

1. In the number 12.345678, what is the name of the place where the 7 is found?
2. What is the name of the 12th place to the right of the decimal point? What would be the value of a 6 in that place?

FUEL FOR THOUGHT

THE INVISIBLE DECIMAL POINT

Professor: Is there a lamp on this table?
Student: Yes.
Professor: Is it on?
Student: No.
Professor (turning the lamp on): Is it on now?
Student: Yes.

Now they look at a whole number, 23.
Professor: Does 23 have a decimal point?
Student: Yes.
Professor: Is it turned on? In other words, can you see it?
Student: No.
Professor: Now look at this number: 23.0. Does it have a decimal point?
Student: Yes.
Professor: Is it turned on?
Student: Yes.
Professor: Like the lamp, the decimal point in a whole number is always there, whether it is turned on or not.

PADDING A DECIMAL WITH ZEROS

There are certain ways to alter a decimal's appearance without changing its value. Padding a decimal with zeros is sometimes a good way to do this.

CAUTION!

HOW NOT TO INSERT ZEROS

BE SURE NOT to insert zeros into a place where the value of the number will be changed. It's okay to write 1.23 as 1.2300. But if you change 1.23 to 1.0023, you will be changing the value of the number.

To the left of the decimal point, add only leading zeros—that is, zeros that come before any non-zero digits.

To the right of the decimal point, add only trailing zeros—zeros that come after any non-zero digits.

On the Left

One way we pad with zeros is to write "0.5" instead of ".5." When you read ".5," the little dot before the 5 is either a fly speck or a decimal point. In contrast, the little dot in "0.5" is almost certainly an actual decimal point.

FUEL FOR THOUGHT

WRITING "0.5" INSTEAD of ".5" is the rule at some hospitals. That rule went into effect after some patients were harmed by doses of medicine that were incorrect by a factor of ten. Decimal points are a serious matter.

When you're working on the left side of the decimal point, you can always add leading zeros. When you change the number 1.23 to 001.23, you have added no tens and no hundreds.

On the Right

Sometimes, it is useful to add zeros to the right of the decimal point. When you change 1.23 to 1.2300, you have added no hundredths and no thousandths. So you haven't changed the value of the number, and you have made a change in appearance that is useful in some situations.

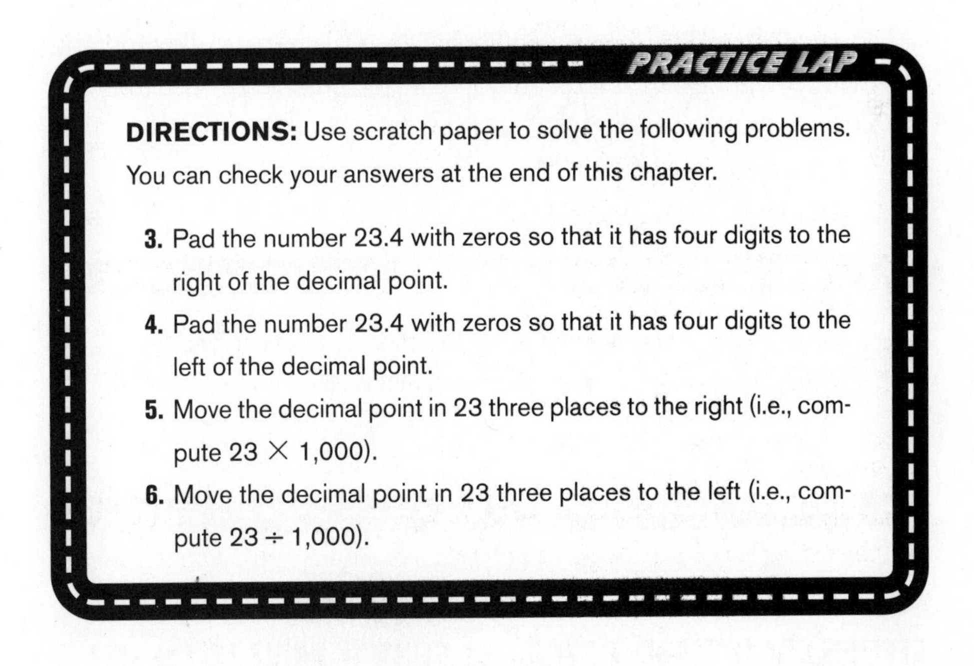

PRACTICE LAP

DIRECTIONS: Use scratch paper to solve the following problems. You can check your answers at the end of this chapter.

3. Pad the number 23.4 with zeros so that it has four digits to the right of the decimal point.
4. Pad the number 23.4 with zeros so that it has four digits to the left of the decimal point.
5. Move the decimal point in 23 three places to the right (i.e., compute $23 \times 1{,}000$).
6. Move the decimal point in 23 three places to the left (i.e., compute $23 \div 1{,}000$).

DIMES TO DOLLARS: MOVING THE DECIMAL POINT TO THE RIGHT

Suppose you have $234.56. You have this money in the form of 2 hundred-dollar bills, 3 tens, 4 one-dollar bills, 5 dimes, and 6 pennies.

I am practicing my (possibly illegal) magic tricks, and I tell you that I will wave my hands over your money, and the following will happen:

Your hundred-dollar bills will turn into thousand-dollar bills.
Your ten-dollar bills will turn into hundred-dollar bills.
Your ones will turn into tens.

Your dimes will turn into dollars.
Your pennies will turn into dimes.

How much money will you have then? You will then have \$2,000 + \$300 + \$40 + \$5 + \$0.60. Your \$234.56 will be transformed into \$2,345.60. You will have ten times as much money as you had before.

That's what happens when you move the decimal point one place to the right.

By shifting the decimal point one place to the right, you multiplied the place value of each digit by ten; therefore, you multiplied the entire number by ten.

Of course, if you were to move a number's decimal point two, three, or four places to the right, you would be multiplying the number by 100; 1,000; or 10,000.

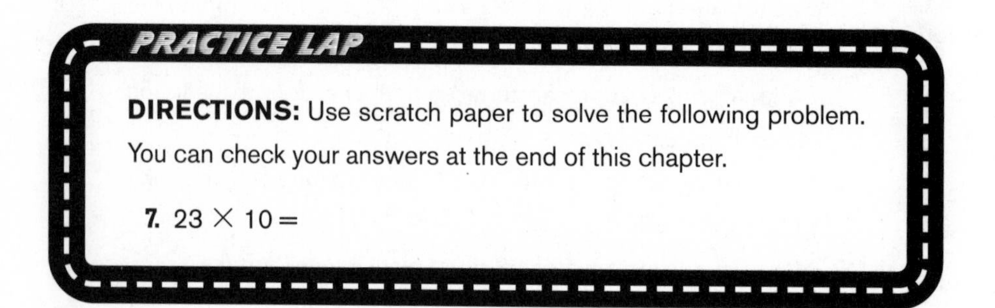

PRACTICE LAP

DIRECTIONS: Use scratch paper to solve the following problem. You can check your answers at the end of this chapter.

7. $23 \times 10 =$

REVERSAL OF FORTUNE: MOVING THE DECIMAL POINT TO THE LEFT

Before you have a chance to examine your new fortune too closely, I wave my hands once again, this time casting a spell on your \$2,345.60.

This time, the opposite transformations happen:

Your thousand-dollar bills change back into hundred-dollar bills.
Your hundred-dollar bills change back into ten-dollar bills.
Your ten-dollar bills change back into ones.
Your one-dollar bills change back into dimes.
Your dimes turn into pennies.

Your \$2,345.60 has turned back into \$234.56. You have only one-tenth as much money as you had a moment ago.

That's what happens when you move the decimal point one place to the left.

Of course, if you move a number's decimal point two, three, or four places to the left, you are multiplying the number by $\frac{1}{100}$, $\frac{1}{1{,}000}$, or $\frac{1}{10{,}000}$.

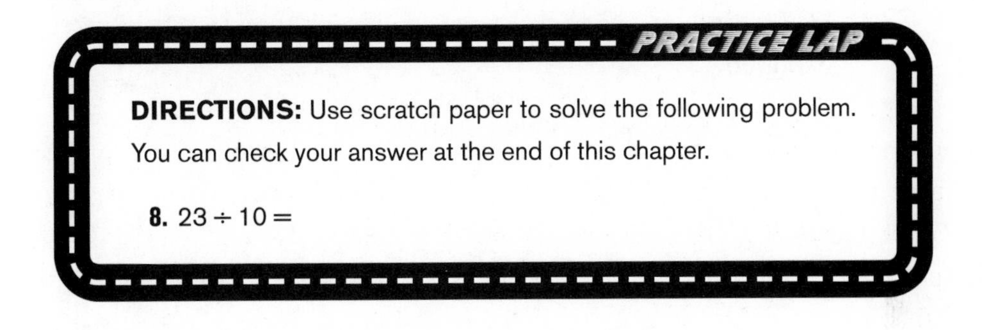

ARRANGING DECIMALS IN ORDER

Line up the decimal points! That is the secret of preparing decimals for ordering, as well as for adding and subtracting.

Example

Arrange these numbers in increasing order:
0.002, 2.000, 46, 23, 0.00098.
Put the numbers under each other, with the decimal points turned on and lined up:

 0.002
 2.000
46.
23.
 0.00098

Now do some zero padding so that every number has the same number of digits:

00.00200
02.00000
46.00000
23.00000
00.00098

You can now arrange these numbers in order just as you would do for whole numbers. Ignore the decimal points to help create the illusion of whole numbers:

00.00098
00.00200
02.00000
23.00000
46.00000

Now drop the padding: 0.00098, 0.002, 2, 23, 46.

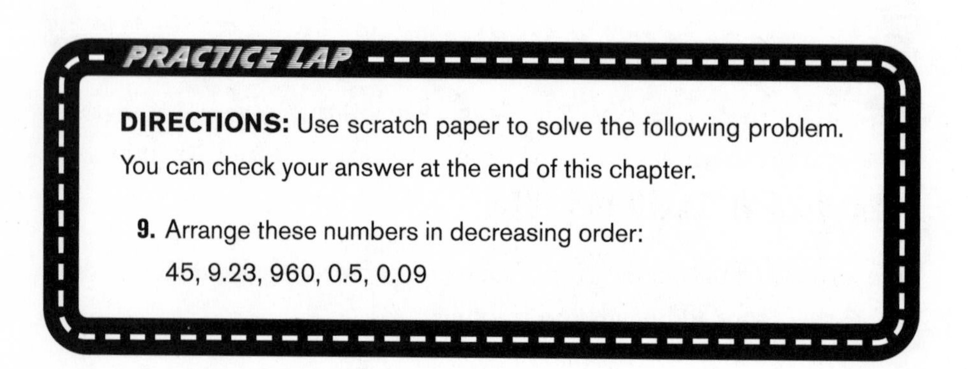

PRACTICE LAP

DIRECTIONS: Use scratch paper to solve the following problem. You can check your answer at the end of this chapter.

9. Arrange these numbers in decreasing order:
45, 9.23, 960, 0.5, 0.09

Adding and Subtracting Decimals

If you spent $4.35 for a burger and $0.75 for a banana, how much did you spend for lunch? That's a decimal addition problem. If you had $27.09 in your pocket before lunch, how much did you have left after lunch? That's a decimal subtraction problem. Adding and subtracting decimals is just everyday math.

When you're adding or subtracting decimals, mathematically speaking, you're carrying out the same operations as when you're adding and subtracting whole numbers. Just keep your columns straight and keep track of where you're placing the decimals in your answers.

Example

$2{,}300 + 0.005 =$

The number 2,300 has a decimal point. It is turned off. Writing "2,300.," the decimal point is now turned on. Lining up the decimal points of your two decimal numbers, you have

$$\begin{array}{r} 2300. \\ +\quad 0.005 \\ \hline 2{,}300.005 \end{array}$$

Example

2,300 – 0.005 =

Lining up the decimal points, you have

$$\begin{array}{r} 2300. \\ -\quad 0.005 \\ \hline \end{array}$$

The blank spaces after the decimal point in 2,300 may be confusing. You can pad 2,300 with zeros after the decimal point: 2,300 = 2,300.000.

Now you have

$$\begin{array}{r} 2{,}300.000 \\ -\quad 0.005 \\ \hline 2{,}299.995 \end{array}$$

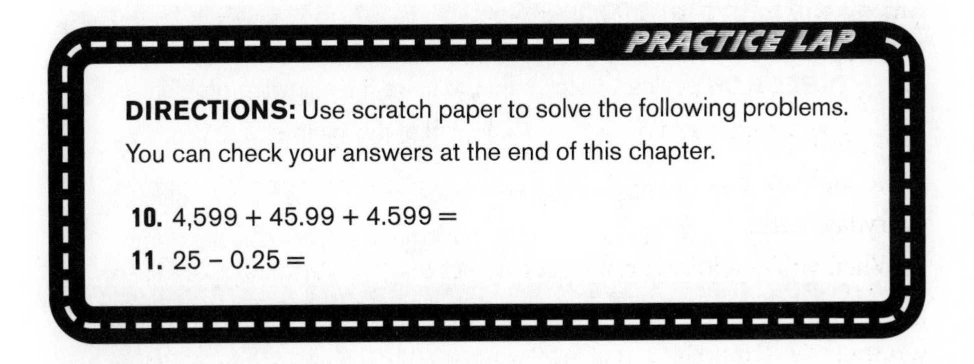

DIRECTIONS: Use scratch paper to solve the following problems. You can check your answers at the end of this chapter.

10. 4,599 + 45.99 + 4.599 =

11. 25 – 0.25 =

Rewriting a Decimal as a Whole Number Times a Power of Ten

You already know how to multiply and divide whole numbers. When it comes to decimals, one approach is to convert them to whole numbers (multiply them by the appropriate power of ten). Then, multiply or divide the whole numbers. Deal separately with the powers of ten.

Example

Write 23.45 as a whole number times a power of ten.

That's easy. You need to move the decimal two places to the right: $23.45 = 2{,}345 \times \frac{1}{100}$.

The fraction $\frac{1}{100}$ is your power of ten. When multiplied by a number (like 2,345), the number $\frac{1}{100}$ shifts the decimal point two places to the left.

Example

Write 2,000 as a one-digit whole number times a power of ten.

$2{,}000 = 2 \times 1{,}000$

In this example, the power of ten, 1,000, shifts the decimal point in the number 2 three places to the right.

Example

Write 0.002 as a one-digit whole number times a power of ten.

$0.002 = 2 \times \frac{1}{1{,}000}$

The power of ten, $\frac{1}{1{,}000}$, shifts the decimal point in the number 2 three places to the left.

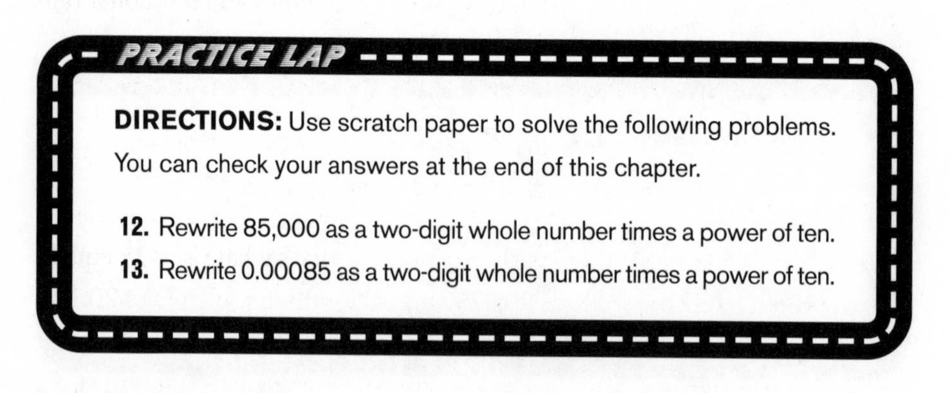
PRACTICE LAP

DIRECTIONS: Use scratch paper to solve the following problems. You can check your answers at the end of this chapter.

12. Rewrite 85,000 as a two-digit whole number times a power of ten.

13. Rewrite 0.00085 as a two-digit whole number times a power of ten.

Multiplying with Decimals

The standard way to multiply two decimals is to multiply the numbers as if they were whole numbers, and then take care of the decimal point.

Example

$8.75 \times 2.3 =$

First, ignore the decimal points: $875 \times 23 = 20{,}125$. Now, take care of the decimal points. The 8.75 has the decimal point two places to the left, and the 2.3 has the decimal point one place to the left. Add: $2 + 1 = 3$. So, you need to move the decimal point of the 20,125 over three places: 20.125.

CAUTION!

A VERY COMMON mistake is putting the decimal place in the wrong place. One shortcut to getting the right answer, while avoiding this mistake, is to do a quick approximation of the answer. For example, if you are multiplying 6.75×9.08, you can use the multiplication fact $6 \times 9 = 54$. This tells you that your answer should be a bit more than 54, letting you rule out 6.129 and 612.900.

What was that all about? Why does this little 2 + 1 thing work so well? Here's why it works: 8.75 is equal to $875 \times \frac{1}{100}$. That's how the place-value system works. The fraction $\frac{1}{100}$ changes dollars to pennies and ten-dollar bills to dimes. It also changes 875 back to 8.75. And the 2.3 is $23 \times \frac{1}{10}$.

So, your multiplication problem, 8.75×2.3, can be written as $875 \times \frac{1}{100} \times 23 \times \frac{1}{10}$. There are four factors: 875, $\frac{1}{100}$, 23, and $\frac{1}{10}$.

Slide the 23 and the 875 next to each other and replace them in the equation with 20,125, because that's what 23×875 equals. Next, slide the $\frac{1}{100}$ next to the $\frac{1}{10}$ and replace them with $\frac{1}{1{,}000}$ because that's what $\frac{1}{100} \times \frac{1}{10}$ equals.

Now you have two numbers, 20,125 and $\frac{1}{1{,}000}$, which multiply to 20.125.

That's what that 2 + 1 operation was all about. You still multiply decimals the way you did before—but now that you understand it, it's not mystery math anymore.

INSIDE TRACK

WHEN YOU MULTIPLY two decimals that are both smaller than 1, your answer is going to be smaller than either of the numbers you multiplied. For 0.1×0.1, your answer is 0.01, which may be read as one one-hundredth.

Example

$2{,}500 \times 0.003$

$2{,}500 = 25 \times 100$ and $0.003 = 3 \times \frac{1}{1{,}000}$.

Multiply the whole numbers: $25 \times 3 = 75$. Then, combine the two powers of ten: $100 \times \frac{1}{1{,}000} = \frac{1}{10}$. The result is $75 \times \frac{1}{10} = 7.5$.

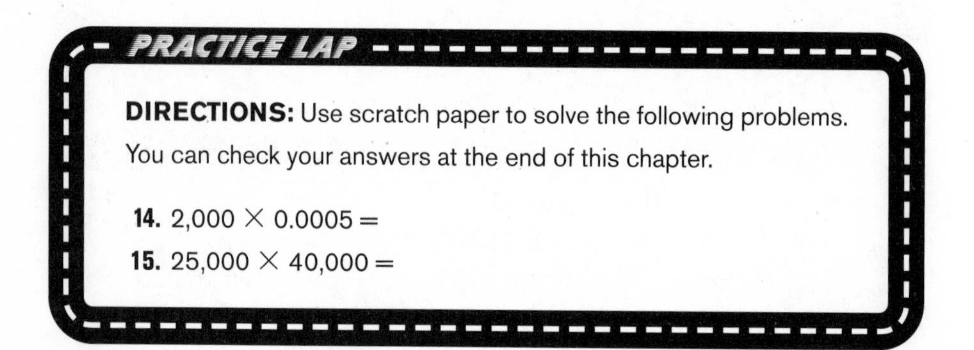

PRACTICE LAP

DIRECTIONS: Use scratch paper to solve the following problems. You can check your answers at the end of this chapter.

14. $2{,}000 \times 0.0005 =$

15. $25{,}000 \times 40{,}000 =$

Dividing in the World of Decimals

One thing to remember when you're dividing one number by another that's less than 1 is that your answer will be larger than the number divided. For example, if you were to divide 4.0 by 0.5, your quotient would be more than 4.0.

For division of decimals, just get rid of the decimals in the divisor (the number by which you divide) and do straight division.

Example

$4.0 \div 0.5$

Let's set it up as a fraction to start: $\frac{4.0}{0.5}$.

Next, move the decimal of the numerator one place to the right, and move the decimal of the denominator one place to the right. Why can you do this? That good old law of arithmetic: Whatever you do to the top you must do to the bottom and vice versa. So, multiply the numerator and the denominator by 10 to get the decimal place moved over one place to the right.

$\frac{4.0}{0.5} \times \frac{10}{10} = \frac{40}{05}$

Then, we do simple division: $\frac{40}{5} = 8$.

You'll notice that 8 (the answer) is larger than 4 (the number divided).

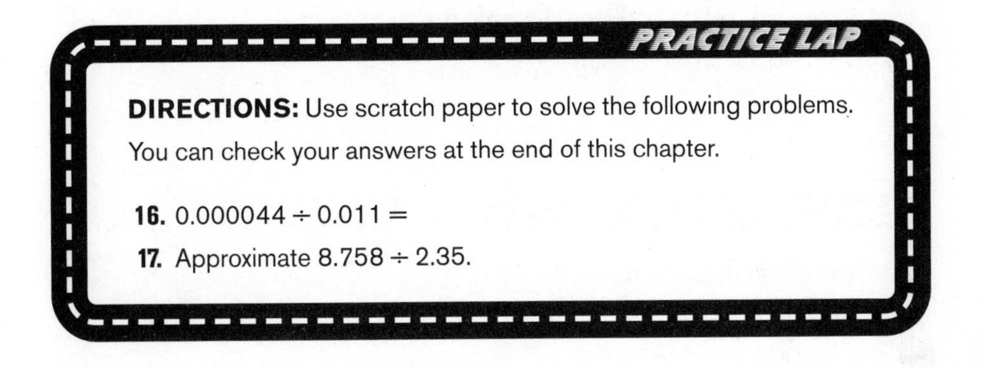

PRACTICE LAP

DIRECTIONS: Use scratch paper to solve the following problems. You can check your answers at the end of this chapter.

16. $0.000044 \div 0.011 =$

17. Approximate $8.758 \div 2.35$.

ANSWERS

1. The 7 is in the fifth place to the right of the decimal point, which is the hundred thousandths place.
2. The name of the 12th place to the right of the decimal point is the trillionths place. The value of a 6 in that place is six trillionths = $\frac{6}{1{,}000{,}000{,}000{,}000} = 0.000000000006$.
3. $23.4 = 23.4000$. You added no tenths, no hundredths, and no thousandths, so you didn't change the value of the number—only the appearance.
4. $23.4 = 0023.4$. Again, you added no hundreds and no thousands, so you didn't change the value of the number—only the appearance.

5. First, turn on the decimal point: 23. Then, pad the number with zeros: 23.000. Now, you are in a position to move the decimal point three places to the right: $23.000 \times 1{,}000 = 23{,}000$.
6. First, turn on the decimal point: 23. Then, pad the number with a zero in front: 023. Now, move the decimal point three places to the left: $023. \div 1{,}000 = 0.023$.
7. $23 \times 10 = 230$. Please don't reach for the calculator for this one. Just move the decimal point one place to the right (after turning the decimal point on and padding with a zero to the right).
8. $23 \div 10 = 2.3$. Again, don't reach for the calculator. Just move the decimal point one place to the left (after turning the decimal point on).
9. Turn on the decimal points for 45 and 960, and write all the numbers under each other with the decimal points lined up.

    45.
    9.23
    960.
    0.5
    0.09

   Do some zero padding:

    045.00
    009.23
    960.00
    000.50
    000.09

   Ignore the decimal points and place the numbers in decreasing order as if they were whole numbers.

    960.00
    045.00
    009.23
    000.50
    000.09

   Remove the zero padding: 960, 45, 9.23, 0.5, 0.09.
10. Turn on the decimal point in 4,599, line up the numbers, pad with zeros, and do the addition without regard to the decimal point column, except to copy the decimal point.

$$\begin{array}{r} 4{,}599.000 \\ 45.990 \\ +\quad 4.599 \\ \hline 4{,}649.589 \end{array}$$

11. Turn on the decimal point in 25, pad some zeros, line up the numbers, and do the subtraction:

$$\begin{array}{r} 25.00 \\ -\quad 0.25 \\ \hline 24.75 \end{array}$$

12. $85{,}000 = 85 \times 1{,}000$
13. $0.00085 = 85 \times 0.00001$
14. $2{,}000 = 2 \times 1{,}000$; $0.0005 = 5 \times 0.0001 = 5 \div 10{,}000$. The original problem becomes $2{,}000 \times 0.0005 = 2 \times 5 \times (1{,}000 \div 10{,}000) = 10 \times (\frac{1}{10}) = 1$. It turns out that the two factors, 2,000 and 0.0005, are reciprocals of each other.
15. $25{,}000 = 25 \times 1{,}000$; $40{,}000 = 40 \times 1{,}000$. $25 \times 40 = 1{,}000$. Altogether, you have $1{,}000 \times 1{,}000 \times 1{,}000 = 1{,}000{,}000{,}000$ (one billion). If you learn a few "equals a thousand" facts, like $25 \times 40 = 20 \times 50 = 10 \times 100 = 5 \times 200$, then you can do this problem and others like it in your head.
16. $0.000044 = 44 \times \frac{1}{1{,}000{,}000}$; $0.011 = 11 \times \frac{1}{1{,}000}$. So, $0.000044 \div 0.011 = (44 \div 11) \times (\frac{1}{1{,}000{,}000} \div \frac{1}{1{,}000}) = 4 \times \frac{1}{1{,}000} = 0.004$.
17. Augment this fraction by a factor of 100 so that the divisor becomes an integer: $8.758 \div 2.35 = 875.8 \div 235$. Now, do the division without regard to the decimal point, which is simply copied to match the one in the dividend (the 875.8).

$$\begin{array}{r} 3.7 \\ 235\overline{)875.8} \\ -705 \\ 1708 \\ -1645 \\ 63 \end{array}$$

The division keeps going at this point, but you can accurately say that $8.758 \div 2.35$ is approximately 3.7.

Percentages—The Basics

WHAT'S AROUND THE BEND

- Converting Percents to Decimals or Fractions
- Some Useful Equivalences
- Finding Percent Changes
- Percentage Distribution
- Finding Percentages of Numbers

Percentages are the mathematical equivalent of fractions and decimals. For example, $\frac{1}{2} = 0.5 = 50\%$. In baseball, a .300 batter is someone who averages 300 base hits every 1,000 times at bat, which is the same as 30 out of 100 ($\frac{30}{100}$ or 30%) or three out of ten ($\frac{3}{10}$). It means the batter gets a hit 30% of the time that he or she comes to bat.

DECIMALS AND PERCENTS

Decimals can be converted into percents by moving their decimal points two places to the right and adding a percent sign. Conversely, percents can be made into decimals by removing the percent sign and moving their decimal points two places to the left.

.45 = 45% .07 = 7% 64% = .64 87% = .87

Now, I'll add a wrinkle. Convert the number 1.2 into a percent. What did you get? Was it 120%? What you do is add a zero to 1.2 and make it 1.20, and then move the decimal two places to the right and add the percent sign. What gives you the right to add a zero? Well, it's okay to do this as long as it does not change the value of the number, 1.2. Because 1.2 = 1.20, you can do this. Can you add a zero to the number 30 without changing its values? Try it. Did you get 300? Does 30 = 300? No—remember that you need to turn on the decimal point first, then add a zero: 30. = 30.0.

Okay, time for another wrinkle. Convert the number 5 into a percent. Did you get 500%? You started with 5, added a decimal point, and a few zeros: 5 = 5.00. Then you converted 5.00 into a percent by moving the decimal point two places to the right and adding a percent sign.

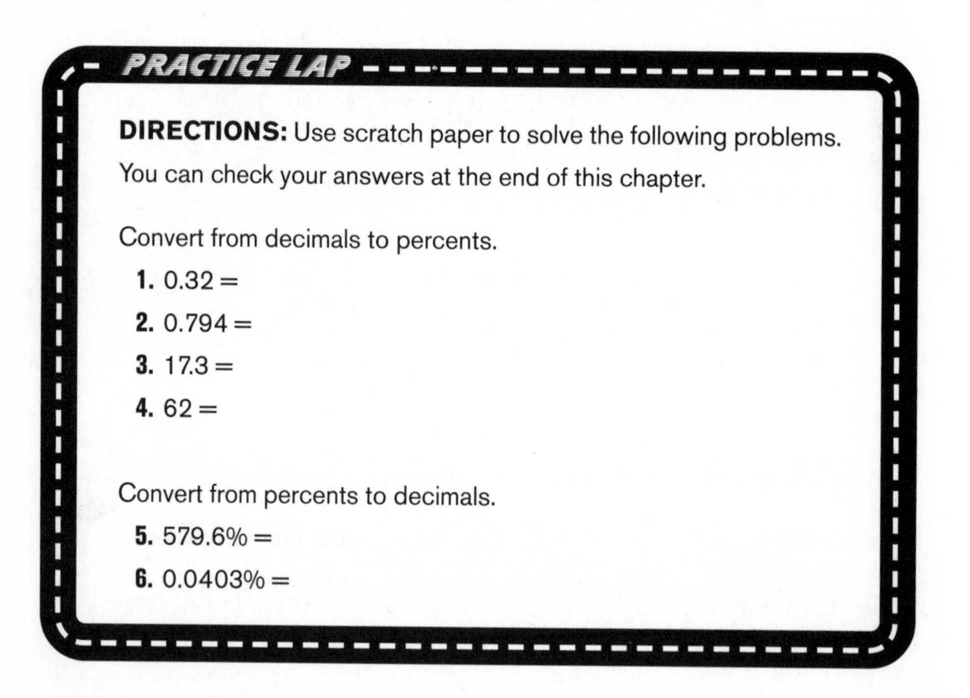

PRACTICE LAP

DIRECTIONS: Use scratch paper to solve the following problems. You can check your answers at the end of this chapter.

Convert from decimals to percents.

1. 0.32 =

2. 0.794 =

3. 17.3 =

4. 62 =

Convert from percents to decimals.

5. 579.6% =

6. 0.0403% =

ENTER FRACTIONS

To change a fraction to a percentage, first change the fraction to a decimal. To do this, divide the numerator by the denominator. Then, change the decimal to a percentage by moving the decimal two places to the right.

$\frac{4}{5} = .80 = 80\%$

$\frac{1}{8} = .125 = 12.5\%$

Now let's try another method for getting from $\frac{1}{4}$ to 25%. You can use the old trick mentioned earlier: Whatever you do to the bottom of a fraction, you must also do to the top. In other words, if you multiply the denominator by a certain number, you must multiply the numerator by the same number.

$$\frac{1}{4} \times \frac{25}{25} = \frac{25}{100}$$

You multiplied both the top and the bottom of the fraction by 25. Why 25? Because you wanted a denominator equal to 100. Having 100 on the bottom of a fraction makes it very easy to convert the fraction into a percent.

All right, you have $\frac{25}{100}$, which comes out to 25%. How did you do this? You removed the 100, or mathematically speaking, you multiplied the fraction by 100, then added a percent sign. In other words,

$$\frac{25}{100} \times \frac{100}{1} = \frac{25}{\cancel{100}_1} \times \frac{\cancel{100}^1}{1} = 25\%$$

Incidentally, *percent* means "per hundred"; 39% means 39 per hundred. To change a percentage to a fraction, divide by 100 and reduce.

$64\% = \frac{64}{100} = \frac{16}{25}$

$82\% = \frac{82}{100} = \frac{41}{50}$

INSIDE TRACK

KEEP IN MIND that any percentage that is 100 or greater will need to reflect a whole number, improper fraction, or mixed number when converted.

$125\% = 1.25, 1\frac{1}{4}, \text{ or } \frac{5}{4}$

$350\% = 3.5, 3\frac{1}{2}, \text{ or } \frac{7}{2}$

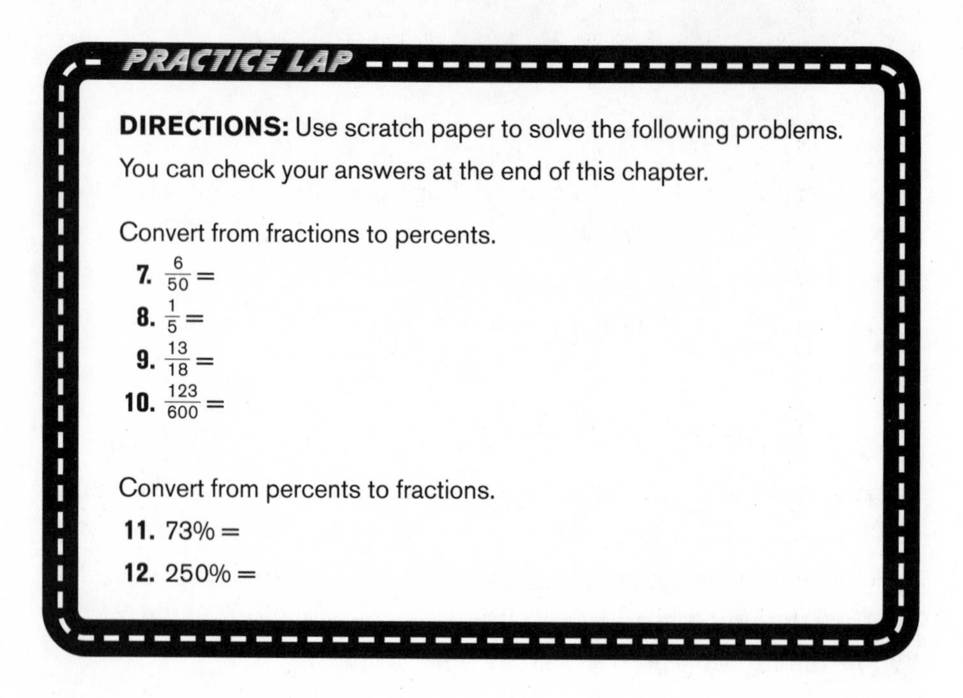
PRACTICE LAP

DIRECTIONS: Use scratch paper to solve the following problems. You can check your answers at the end of this chapter.

Convert from fractions to percents.

7. $\frac{6}{50} =$

8. $\frac{1}{5} =$

9. $\frac{13}{18} =$

10. $\frac{123}{600} =$

Convert from percents to fractions.

11. 73% =

12. 250% =

Some Useful Equivalences

While we are doing conversions, here are some fraction-percent equivalences. It is good to learn these equivalences by heart, but it will take some time. You might consider making a set of flash cards with these equivalences, which you can study on the go.

$\frac{1}{2} = 50\%$

$\frac{1}{3} = 33\frac{1}{3}\%$ (approximately equal to 33%)

$\frac{2}{3} = 66\frac{2}{3}\%$ (approximately equal to 67%)

$\frac{1}{4} = 25\%$

$\frac{2}{4} = \frac{1}{2} = 50\%$

$\frac{3}{4} = 75\%$

$\frac{1}{5} = 20\%$

$\frac{2}{5} = 40\%$

$\frac{3}{5} = 60\%$

$\frac{4}{5} = 80\%$

$\frac{1}{6}$ is approximately equal to 17%

$\frac{2}{6} = \frac{1}{3}$ is approximately equal to 33%

$\frac{3}{6} = \frac{1}{2} = 50\%$

$\frac{4}{6} = \frac{2}{3}$ is approximately equal to 67%

$\frac{5}{6}$ is approximately equal to 83%

$\frac{1}{7}$ is approximately equal to 14%

$\frac{6}{7}$ is approximately equal to 86%

$\frac{1}{8} = 12.5\%$ (half of 25%)

$\frac{2}{8} = \frac{1}{4} = 25\%$

$\frac{3}{8} = 37.5\%$ ($\frac{3}{8} = \frac{1}{8} + \frac{2}{8}$)

$\frac{4}{8} = \frac{2}{4} = \frac{1}{2} = 50\%$

$\frac{5}{8} = 62.5\%$ ($\frac{5}{8} = \frac{4}{8} + \frac{1}{8}$)

$\frac{6}{8} = \frac{3}{4} = 75\%$

$\frac{7}{8} = 87.5\%$ ($\frac{7}{8} = 1 - \frac{1}{8}$)

$\frac{1}{9}$ is approximately equal to 11%

$\frac{2}{9}$ is approximately equal to 22%

$\frac{3}{9} = \frac{1}{3}$ is approximately equal to 33%

$\frac{4}{9}$ is approximately equal to 44%

$\frac{5}{9}$ is approximately equal to 56%

$\frac{6}{9} = \frac{2}{3}$ (approximately equal to 67%)

$\frac{7}{9}$ is approximately equal to 78%

$\frac{8}{9}$ is approximately equal to 89%

$\frac{1}{10} = 10\%$

$\frac{2}{10} = \frac{1}{5} = 20\%$

$\frac{3}{10} = 30\%$

$\frac{4}{10} = \frac{2}{5} = 40\%$

$\frac{5}{10} = \frac{1}{2} = 50\%$

$\frac{6}{10} = \frac{3}{5} = 60\%$

$\frac{7}{10} = 70\%$

$\frac{8}{10} = \frac{4}{5} = 80\%$

$\frac{9}{10} = 90\%$

FINDING PERCENTAGE CHANGES

Imagine you were earning $500 and got a $20 raise. By what percentage did your salary go up? Try to figure it out.

We have a nice formula to help us solve percentage change problems. Here's how it works: Your salary is $500, so that's the original number. You got a $20 raise; that's the change. The formula looks like this:

$$\text{percentage change} = \frac{\text{change}}{\text{original number}}$$

Next, you substitute the numbers into the formula. Then solve it. Once you have $\frac{\$20}{\$500}$, you could reduce it all the way down to $\frac{1}{25}$ and solve it using division:

$$= \frac{\$20}{\$500} = \frac{2}{50} = \frac{4}{100} = 4\%$$

$$25\overline{)1.00} = 4\%$$

(quotient: .04)

−1.00

Example

On New Year's Eve, you made a resolution to lose 30 pounds by the end of the year. And sure enough, your weight dropped from 140 pounds to 110. By what percentage did your weight fall?

$$\text{percentage change} = \frac{\text{change}}{\text{original number}} = \frac{30}{140} = \frac{3}{14}$$

$$14\overline{)3.0000} = 21.4\%$$

(quotient: .2142)

− 2 8
20
− 14
60
− 56
40
− 28

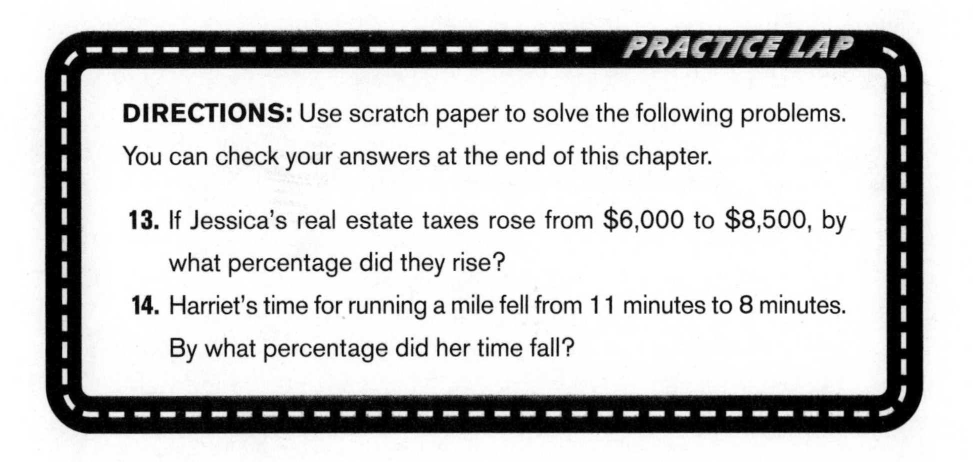

PRACTICE LAP

DIRECTIONS: Use scratch paper to solve the following problems. You can check your answers at the end of this chapter.

13. If Jessica's real estate taxes rose from $6,000 to $8,500, by what percentage did they rise?

14. Harriet's time for running a mile fell from 11 minutes to 8 minutes. By what percentage did her time fall?

Percentage Increases

Pick a number, any number. Now triple it. By what percentage did this number increase? Take your time. Use the space in the margin to calculate the percentage.

What did you get? 300%? Nice try, but I'm afraid that's not the right answer.

I'm going to pick a number for you, and then you triple it. I pick the number 100. Now I'd like you to use the percentage change formula to get the answer. (Incidentally, you may have gotten the right answer, so you may be wondering why I'm making such a big deal. But I know from sad experience that almost no one gets this right on the first try.)

So where were we? The formula. Write it down in the space below, substitute numbers into it, and then solve it.

Let's go over this problem step by step. We picked a number, 100. Next, we tripled it, which gives us 300. Right? Now we plug some numbers into the formula. Our original number is 100. And the change when we go from 100 to 300? It's 200. From there, it's just arithmetic: 200 ÷ 100 = 2, or 200%. (Remember, to find the percent, move the decimal two places to the right.)

$$\text{Percentage change} = \frac{\text{change}}{\text{original number}} = \frac{200}{100} = 2.0 = 200\%$$

This really isn't that hard. In fact, you're going to get really good at just looking at a couple of numbers and figuring out percentage changes in your head. Whenever you go from 100 to a higher number, the percentage increase is the difference between 100 and the new number. Suppose you were to quadruple a number. What's the percentage increase? It's 300% (400 – 100). When you double a number, what's the percentage increase? It's 100% (200 – 100).

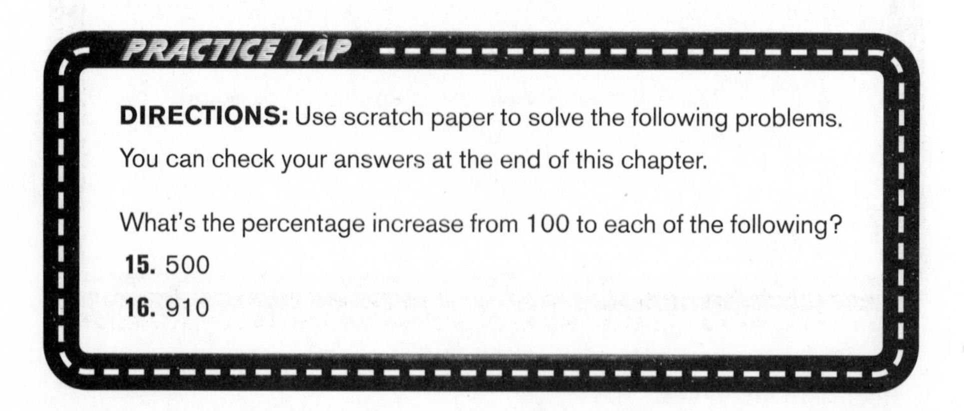

PRACTICE LAP

DIRECTIONS: Use scratch paper to solve the following problems. You can check your answers at the end of this chapter.

What's the percentage increase from 100 to each of the following?

15. 500

16. 910

The number 100 is very easy to work with. Sometimes, you can use it as a substitute for another number. For example, what's the percentage increase if we go from 3 to 6? Isn't it the same as if you went from 100 to 200? It's a 100% increase.

What's the percentage increase from 5 to 20? It's the same as the one from 100 to 400. It is a 300% increase.

What we've been doing here is just playing around with numbers, seeing if we can get them to work for us. As you get more comfortable with numbers, you can try to manipulate them the way we just did.

Percentage Decreases

Remember the saying "whatever goes up must come down"? If Melanie was earning $100 and her salary was cut to $93, by what percent was her salary cut?

You might know intuitively that the answer is 7%. More formally, we divided the change in salary, $7, by the original salary, $100:

$$\frac{\$7}{\$100} = 7\%$$

What would be the percentage decrease from 100 to 10?

$$\frac{90}{100} = 90\%$$

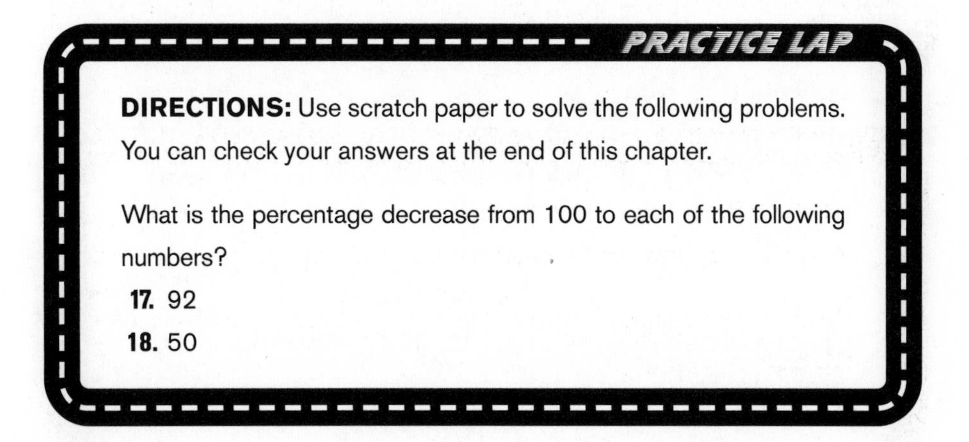

PRACTICE LAP

DIRECTIONS: Use scratch paper to solve the following problems. You can check your answers at the end of this chapter.

What is the percentage decrease from 100 to each of the following numbers?

17. 92

18. 50

Now I'm going to throw you another curve ball. If a number—any number—were to decline by 100%, what number would you be left with? I'd really like you to think about this one.

What did you get? You should have gotten 0. That's right—no matter what number you started with, a 100% decline leaves you with 0.

Percentage Distribution

Percentage distribution tells you the number per hundred that is represented by each group in a larger whole. For example, in Canada, 30% of the people live in cities, 45% live in suburbs, and 25% live out in the country. When you calculate percentage distributions, you'll find that they always add up to 100% (or a number very close to 100, depending on the exact decimals involved). If they don't, you'll know that you have to redo your calculations.

A class had half girls and half boys. What percentage of the class was girls, and what percentage of the class was boys? The answers are obviously 50% and 50%. That's all there is to percentage distribution. Of course, the problems do get a bit more complicated, but all percentage distribution problems start out with one simple fact: The distribution will always add up to 100%.

Here's another one. One-quarter of the players on a baseball team are pitchers, one-quarter are outfielders, and the rest are infielders. What is the team's percentage distribution of pitchers, infielders, and outfielders?

Pitchers are $\frac{1}{4}$, or 25%; outfielders are also $\frac{1}{4}$, or 25%. So infielders must be the remaining 50%.

INSIDE TRACK

WHEN DOING PERCENTAGE distribution problems, it's always a good idea to check your work. If your percentages don't add up to 100 (or 99 or 101), then you've definitely made a mistake, so you'll need to go back over all your calculations. Because of rounding, you'll occasionally end up with 100.1 or 99.9 when you check, which is fine.

FINDING PERCENTAGES OF NUMBERS

The Internal Revenue Service charges different tax rates for different levels of income. For example, most middle-income families are taxed at a rate of

28% on some of their income. Suppose that one family had to pay 28% of $10,000. How much would that family pay?

$10,000 × .28 = $2,800

You'll notice that we converted 28% into the decimal .28 to carry out that calculation.

Example

How much is 14.5% of 1,304?

$$\begin{array}{r} 1{,}304 \\ \times\ .145 \\ \hline 6520 \\ 5216 \\ +1304 \\ \hline 189.080 \end{array}$$

Most people try to leave a 15% tip in restaurants. In New York City, where the sales tax is 8.65%, customers often just double the tax. But there is actually another very fast and easy way to calculate that 15% tip.

Let's say that your bill comes to $28.19. Round it off to $28, the nearest even dollar amount. Then find 10% of $28, which is $2.80. Now what's half of $2.80? It's $1.40. How much is $1.40 plus $2.80? It's $4.20.

Let's try a much bigger check—$131.29. Round it off to the nearest even dollar amount—$132. What is 10% of $132? It's $13.20. And how much is half of $13.20? It's $6.60. Finally, add $13.20 and $6.60 together to get your $19.80 tip.

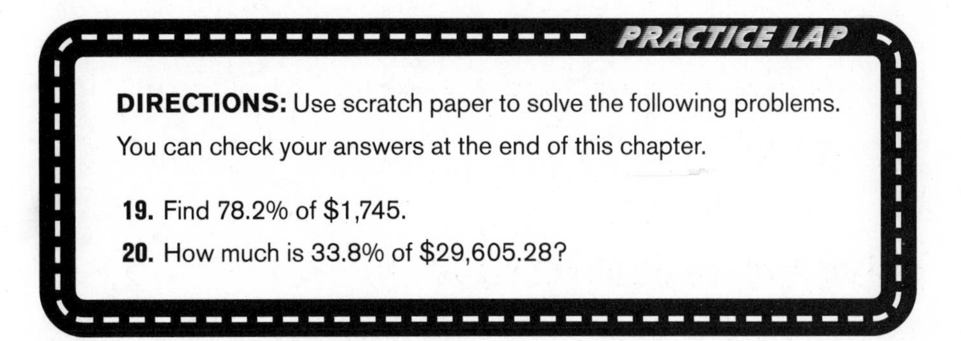

PRACTICE LAP

DIRECTIONS: Use scratch paper to solve the following problems. You can check your answers at the end of this chapter.

19. Find 78.2% of $1,745.

20. How much is 33.8% of $29,605.28?

ANSWERS

1. $0.32 = 32\%$
2. $0.794 = 79.4\%$
3. $17.3 = 1{,}730\%$
4. $62 = 6{,}200\%$
5. $579.6\% = 5.796$
6. $0.0403\% = 0.000403$
7. $\frac{6}{50} = \frac{6}{50} \times \frac{2}{2} = \frac{12}{100} = 12\%$
8. $\frac{1}{5} = \frac{1}{5} \times \frac{20}{20} = \frac{20}{100} = 20\%$
9. $\frac{13}{18} = 18\overline{)13.000}$ = 72.2% (quotient .722)

   $-12\,6$
   40
   -36
   40
   -36

10. $\frac{123}{600} = 600\overline{)123.000} = 6\overline{)1.230} = 20.5\%$ (quotient .205)
11. $73\% = \frac{73}{100}$
12. $250\% = \frac{250}{100} = 2\frac{1}{2}$
13. $\text{percentage change} = \frac{\text{change}}{\text{original number}} = \frac{\$2{,}500}{\$6{,}000} = \frac{25}{60} = \frac{5}{12}$

   $12\overline{)5.0000} = 41.7\%$ (quotient .4166)

   $-4\,8$
   20
   -12
   80
   -72
   80
   -72

14. percentage change = $\frac{\text{change}}{\text{original number}} = \frac{3}{11}$

$$\begin{array}{r} .2727 \\ 11\overline{)3.0000} = 27.3\% \\ -2\,2 \\ 80 \\ -77 \\ 30 \\ -22 \\ 80 \\ -77 \end{array}$$

15. $\frac{400}{100} = 400\%$

16. $\frac{810}{100} = 810\%$

17. $\frac{8}{100} = 8\%$

18. $\frac{50}{100} = 50\%$

19.

$$\begin{array}{r} \$1{,}745 \\ \times\ .782 \\ \hline 3\,490 \\ 139\,60 \\ +\ 1221\,5 \\ \hline \$1{,}364.59\not{0} \end{array}$$

20.

$$\begin{array}{r} \$29{,}605.28 \\ \times\ .338 \\ \hline 23684224 \\ 8881584 \\ +\ \ 8881584 \\ \hline \$10{,}006.58\not{4}\not{6}\not{4} \end{array}$$

7 The Saga of Signed Numbers

WHAT'S AROUND THE BEND

- Negative Numbers
- Signed Numbers
- Opposite Numbers
- Finding the Absolute Value of a Signed Number
- Arranging Signed Numbers in Order
- Adding Signed Numbers
- Subtracting Signed Numbers
- Multiplying Signed Numbers
- Dividing Signed Numbers

NEGATIVE NUMBERS

Negative numbers show up on the number line to the left of the zero. They are the numbers like -1, -2, and so forth, as well as the negative numbers not shown on the number line, like $-\frac{1}{10}$ and -3.8.

The Number Line

<——|——|——|——|——|——|——|——|——|——|——|——|——|——|——|——>

−7 −6 −5 −4 −3 −2 −1 0 1 2 3 4 5 6 7

You encounter negative numbers when the temperature goes below zero, when debts exceed assets so that net financial worth becomes negative, or in business when costs exceed revenue so that profit is negative, indicating a loss.

Negative numbers fit nicely into the scheme of things. The methods you are about to review for negative numbers have worked very well for the kinds of numbers you have already reviewed: whole numbers, fractions, mixed numbers, and decimals.

SIGNED NUMBERS

You use the expression *signed numbers* when you are talking about numbers that can be either positive or negative.

In the next sections, you will review the manner in which signed numbers pair off with each other into opposites. You will compare signed numbers with each other, and you will review the four basic operations (addition, subtraction, multiplication, and division), as applied to signed numbers.

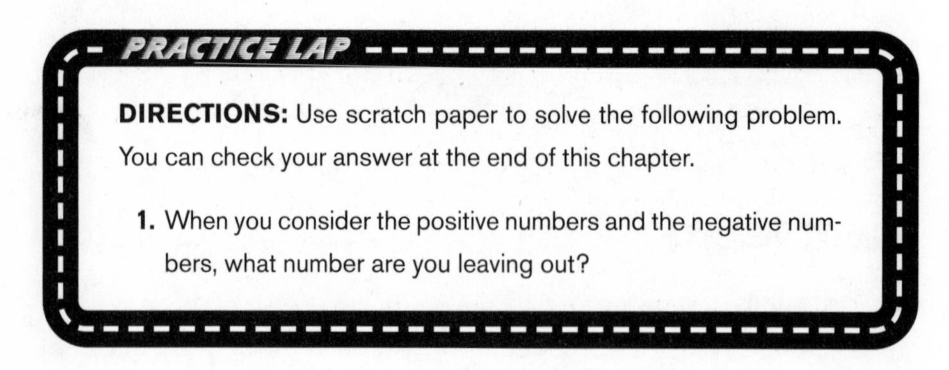

PRACTICE LAP

DIRECTIONS: Use scratch paper to solve the following problem. You can check your answer at the end of this chapter.

1. When you consider the positive numbers and the negative numbers, what number are you leaving out?

PAIRS OF OPPOSITE NUMBERS

The numbers 3 and –3 are opposites of each other. To be more specific, 3 and –3 are **additive** opposites of each other.

So, –3 is the opposite of 3, and 3 is the opposite of –3. On a number line, 3 and –3 are on opposite sides of the zero.

The additive opposite of a number is the mirror image of that number through zero on the number line.

Suppose you were to place a double-sided mirror, upright, on the zero of the number line. Looking at the mirror from the right-hand side, the 3 would appear in the mirror to the left of the 0, at the place where negative 3 really is. And looking from the left, –3 would appear in the mirror to the right of the 0. The –3 would appear to be where the 3 really is.

That's why you say that 3 and –3 are mirror images of each other. Pairs of opposite numbers are mirror images of each other.

FUEL FOR THOUGHT

YOU ALSO SAY 3 and –3 are *negatives* of each other. It is correct to say that 3 is the opposite of –3. It is equally correct to say that 3 is the *negative* of –3.

Why Pairs of Opposites Are Important

Pairs of opposite numbers, or mirror-image numbers, are important because when you add a pair together, you get zero. For instance, $3 + (-3) = 0$.

Suppose I add 3 to 17 ($17 + 3 = 20$). Now, I want my 17 back. I can add –3 to 20 to get 17.

$$20 + (-3) = 17$$

I got my 17 back. I added 3 to 17, and then I added the opposite of 3. The effect was that adding –3 undid the act of adding 3.

FUEL FOR THOUGHT

WHEN YOU BALANCE your checkbook, you are working with positive and negative numbers. Deposits are positive numbers. Checks, withdrawals, and service charges are negative numbers. So, balancing your checkbook lets you practice using both positive and negative numbers.

Does this work in the other order? Suppose I add –3 to 17: $17 + (-3) = 14$. Now I add 3: $14 + 3 = 17$. There's my 17 back. So, it does work in that order. Adding 3 undid the effect of adding –3, and adding –3 undid the effect of adding 3. This ability to do and then undo comes in very handy as you get deeper into math.

How Do You Flip a Number into Its Opposite?

You put a hyphen in front of it.

Okay, you understand that. But mathematically, how do you turn a number into its opposite? You multiply it by –1. Let's look at two examples.

Examples

Multiply 3 by –1 to get –3, like this: $-1 \times 3 = -3$.

Multiply –3 by –1 to get 3: $-1(-3) = -(-3) = 3$.

So multiplying by –1 will flip a positive number into its opposite, which is negative, and multiplying by –1 will flip a negative number into its opposite, which is positive. Multiplying by –1 will flip a number either way.

FINDING THE ABSOLUTE VALUE OF A SIGNED NUMBER

The **absolute value** of a number, sometimes called the **magnitude** of the number, is the distance between the number and the zero of the number line. For example, the number 3 is three units from the zero, so the **absolute value** of 3 is 3. The number –3 is also three units from the zero, so the **absolute value** of –3 is also 3. Absolute value is always positive.

The symbol for absolute value is two straight lines surrounding an expression, for example, $|-4|$ $|137|$.

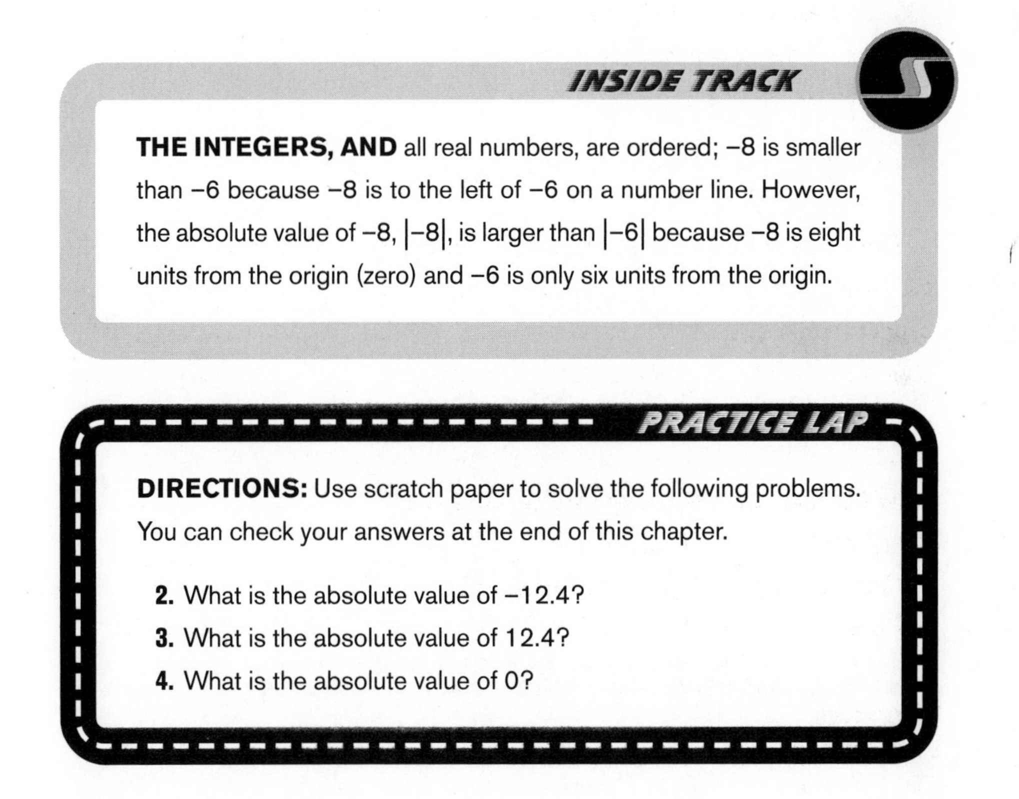

INSIDE TRACK

THE INTEGERS, AND all real numbers, are ordered; –8 is smaller than –6 because –8 is to the left of –6 on a number line. However, the absolute value of –8, $|-8|$, is larger than $|-6|$ because –8 is eight units from the origin (zero) and –6 is only six units from the origin.

PRACTICE LAP

DIRECTIONS: Use scratch paper to solve the following problems. You can check your answers at the end of this chapter.

2. What is the absolute value of –12.4?

3. What is the absolute value of 12.4?

4. What is the absolute value of 0?

ARRANGING SIGNED NUMBERS IN ORDER

Less than means "to the left of" on the number line.

You say $3 < 5$ (read "3 is less than 5") or, with the exact same meaning, $5 > 3$ (read "5 is greater than 3").

How do you know that 3 is less than 5 ($3 < 5$)? A glance at the number line confirms that 3 is to the left of 5.

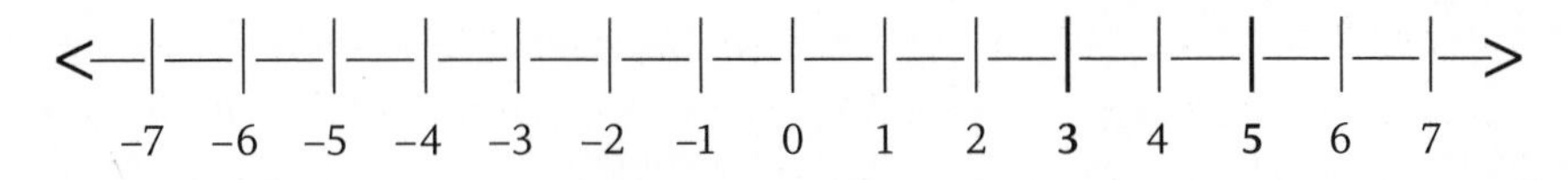

As for negative numbers, an equally quick glace at the number line will show you that –3 is to the *right* of –5 or, in other words, –5 is to the left of –3: –5 < –3.

<—|—|—|—|—|—|—|—|—|—|—|—|—|—|—|—>

–7 –6 **–5** –4 **–3** –2 –1 0 1 2 3 4 5 6 7

So if I give you two numbers, how do you know which is the lesser one? Well, if they are both positive, the one that is "less than" is the smaller one.

And what if they are both negative? Then the one that is further way from zero of the number line is the one that is "less than." That's why –5 is less than –3 (–5 < –3).

This means that if both numbers are negative, the one with the larger absolute value is the one that is smaller.

One more question: What if the two numbers you are comparing have opposite signs?

That's easy. Take one more glance at the number line. You will see that every negative number is to the left of every positive number. Looking at –5 and 3, the –5 is to the left of 3, so –5 < 3.

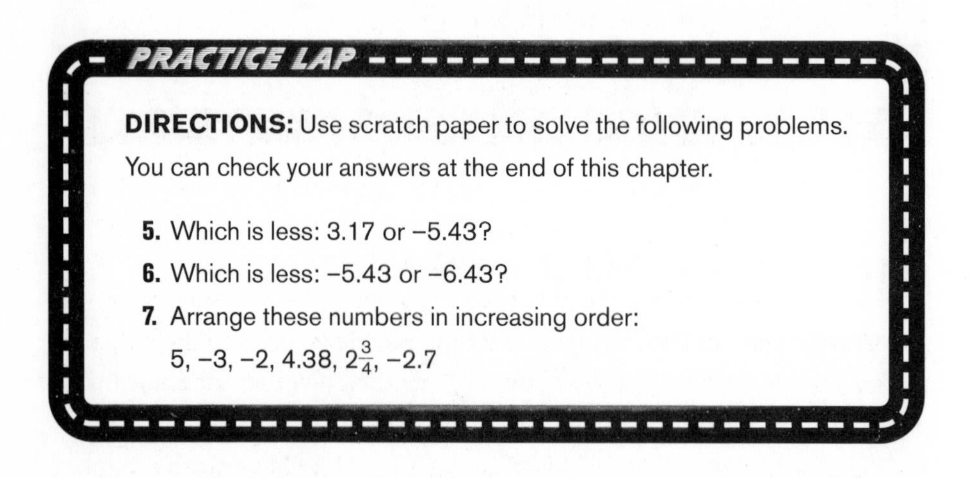

PRACTICE LAP

DIRECTIONS: Use scratch paper to solve the following problems. You can check your answers at the end of this chapter.

5. Which is less: 3.17 or –5.43?

6. Which is less: –5.43 or –6.43?

7. Arrange these numbers in increasing order:
5, –3, –2, 4.38, $2\frac{3}{4}$, –2.7

Adding Signed Numbers

You can understand the addition of integers, or any signed numbers, by using a number line. When you add a positive number, you move to the right, and when you add a negative integer, you move to the left. The answer to the problem 5 + 2 is 7, because you start at the origin, move five units to the right, and then move two more units right, ending on +7.

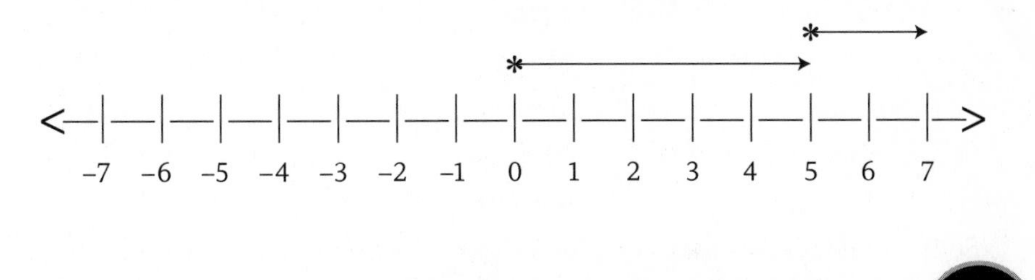

FUEL FOR THOUGHT

Integers are the set of whole numbers and their opposites–that is –4, –3, –2, –1, 0, 1, 2, 3, 4, . . .

Integers can be shown on a number line, where 0 is also called the **origin**.

The answer to the problem –3 + –2 is –5, because you start at the origin, move three units to the left, and then move two more units to the left, ending on –5.

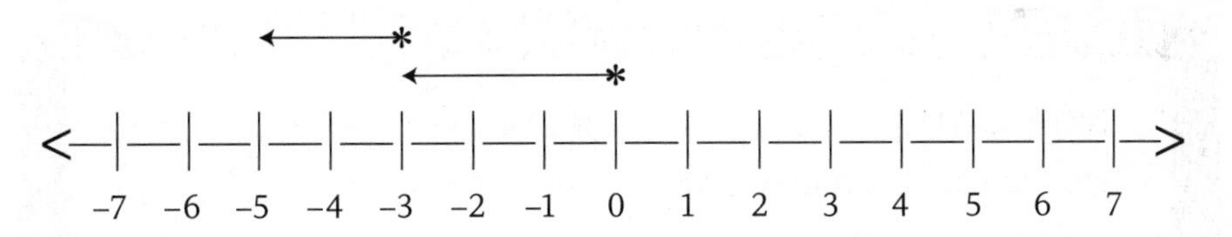

Now, look at 7 + –2. The answer to this problem is 5—you start at the origin, move seven units to the right, and then move two units to the left, ending on +5.

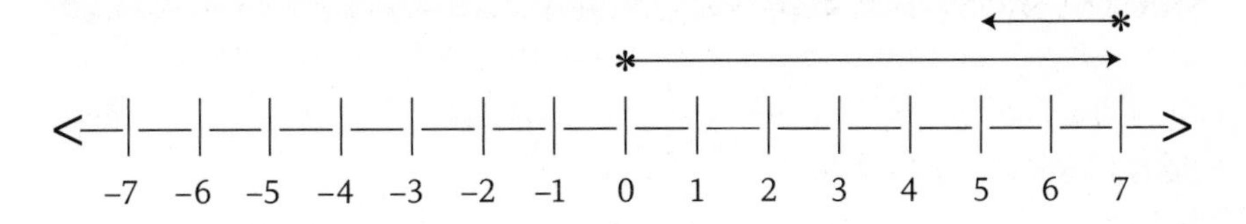

Notice that adding two numbers with the same sign yields the sum of the absolute values, and the sign stays the same. Adding two numbers with different signs yields the difference between the absolute values, and the sign of the number with the larger absolute value will dictate the sign of the answer. You actually subtract absolute values when you add two numbers with different signs. This leads to the special case when you add two opposites, which always results in an answer of zero:

$-7 + 7 = 0$

$3 + -3 = 0$

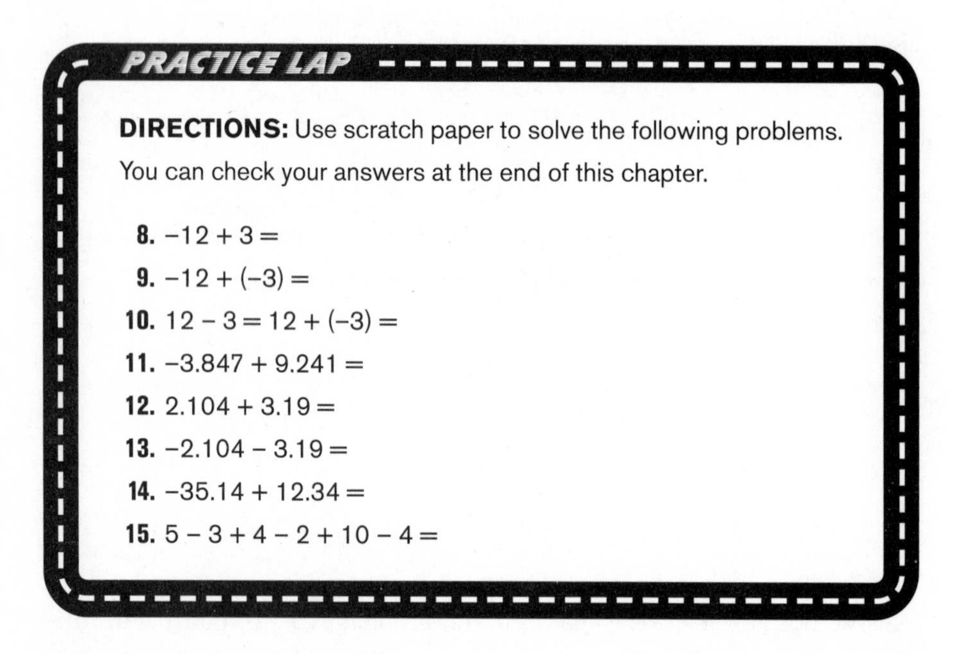

PRACTICE LAP

DIRECTIONS: Use scratch paper to solve the following problems. You can check your answers at the end of this chapter.

8. $-12 + 3 =$

9. $-12 + (-3) =$

10. $12 - 3 = 12 + (-3) =$

11. $-3.847 + 9.241 =$

12. $2.104 + 3.19 =$

13. $-2.104 - 3.19 =$

14. $-35.14 + 12.34 =$

15. $5 - 3 + 4 - 2 + 10 - 4 =$

Subtracting Signed Numbers

It is best to think of the subtracting of signed numbers as adding the opposite of the second number. For example, if the problem states that $-10 - 6$, you would change this problem to $-10 + -6$, and follow the rules of adding signed numbers.

INSIDE TRACK

WHEN YOU SEE a double negative, immediately change the double negative to one positive sign. For example, 6 – –8 simplifies to 6 + 8.

USE THE FOLLOWING flowchart to learn the rules for adding and subtracting signed numbers.

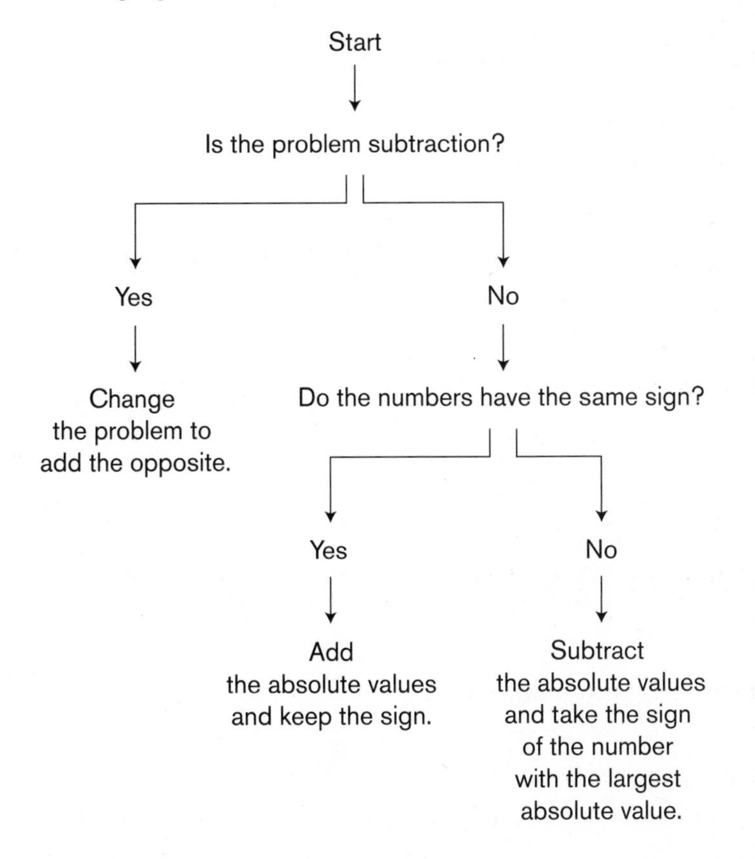

For example, –7 – 4 is subtraction, so change it to –7 + –4 by adding the opposite. The numbers have the same sign, so add the absolute values and keep the sign to arrive at –11.

To add or subtract a string of signed numbers, you can pick up pairs of numbers, working from left to right. Then, follow the rules given for each pair. For example, to evaluate –5 + 7 – 3, first evaluate, –5 + 7. By following the aforementioned flowchart, you are adding two numbers with different signs. Subtract the absolute values and take the sign of the number with the larger absolute value, resulting in +2. Then evaluate this number with the next number in the string: +2 – 3. Subtract 2 from 3, and take the sign of the –3, because –3 has a larger absolute value. Your answer is –1.

PRACTICE LAP

DIRECTIONS: Use scratch paper to solve the following problems. You can check your answers at the end of this chapter.

16. 8 – 5 =
17. 5 – 8 =
18. –3.56 – (–8.62) =
19. 12.8 – (–4) =

INSIDE TRACK

IF YOUR CALCULATOR has the positive/negative key, then it will perform positive and negative arithmetic. Most calculators have you enter the sign of a negative number after you enter the absolute value. So, –8 would be entered as "8" and "+/–." To perform –3 – –6, you would use the key sequence: "3," "+/–," "–," "6," "+/–," "=."

Multiplying Signed Numbers

How much is (–3) × (–5)? Here are the rules for multiplying signed numbers:

positive × positive = positive
positive × negative = negative
negative × positive = negative
negative × negative = positive

So, the answer to $(-3) \times (-5)$ is $(-3) \times (-5) = 15$.

Suppose you multiply a whole string of numbers, like this:

$(-3) \times (5) \times (2) \times (-1) \times (2) \times (-2) = ?$

You can see how this will work by doing the multiplications one at a time.

$(-3) \times (5) = -15$

$-15 \times 2 = -30$

$-30 \times -1 = 30$

$30 \times 2 = 60$

$60 \times -2 = -120$

You see that each time you encounter a negative number, the sign of the result changes. So, if there is an even number of negatives, the final sign will be plus. If there are an odd number of negatives, the sign will be negative.

In the previous example, there are three negative numbers. Three is odd. So the result is a negative number. It's like flipping a light switch on and off. Each time you change the position of the switch, the light changes from on to off, or from off to on. If you start with "off," and flip the switch an even number of times, the result will be "off." If you flip the switch an odd number of times, the result will be "on."

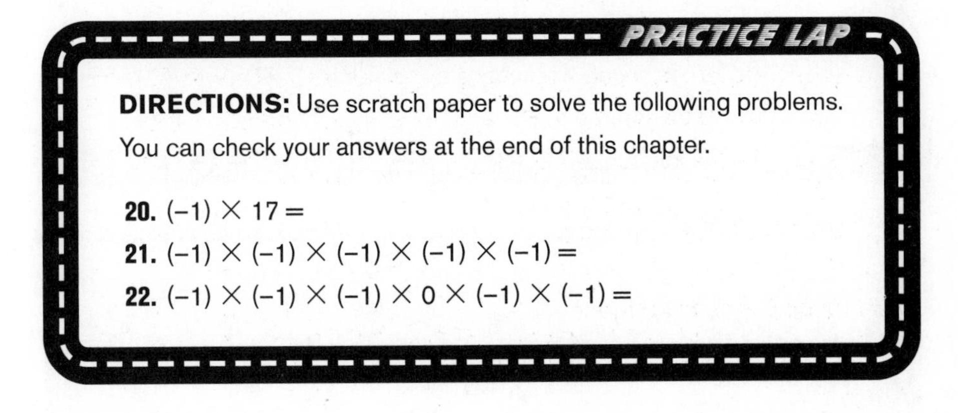

PRACTICE LAP

DIRECTIONS: Use scratch paper to solve the following problems. You can check your answers at the end of this chapter.

20. $(-1) \times 17 =$

21. $(-1) \times (-1) \times (-1) \times (-1) \times (-1) =$

22. $(-1) \times (-1) \times (-1) \times 0 \times (-1) \times (-1) =$

Dividing Signed Numbers

Dividing works just like multiplying as far as signs are concerned.

positive ÷ positive = positive
positive ÷ negative = negative
negative ÷ positive = negative
negative ÷ negative = positive

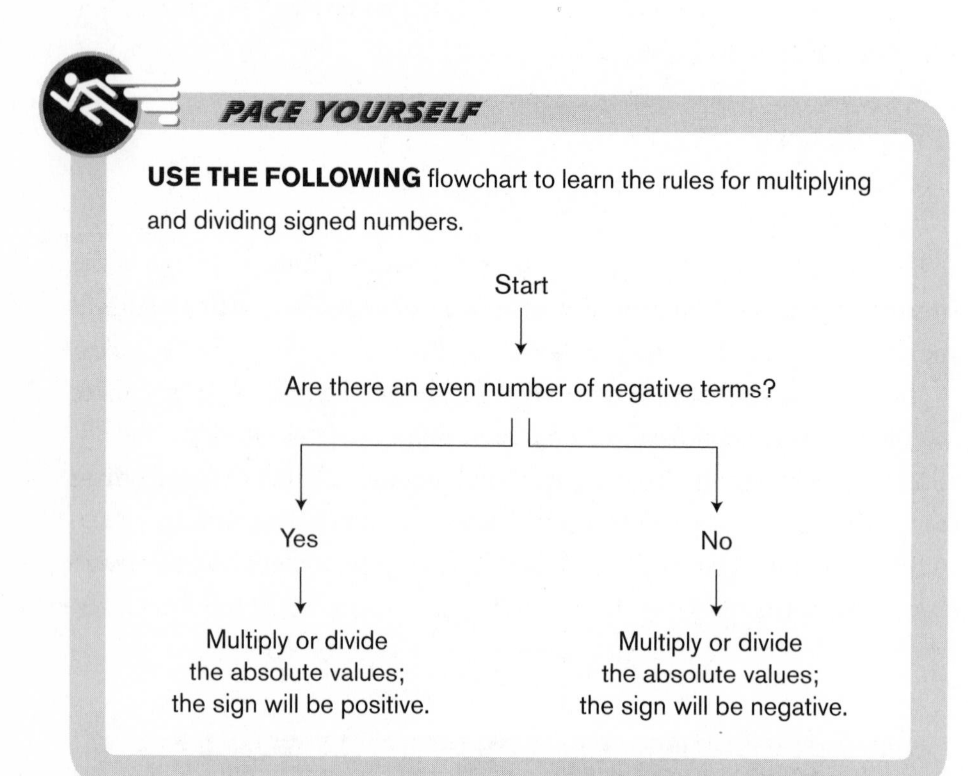

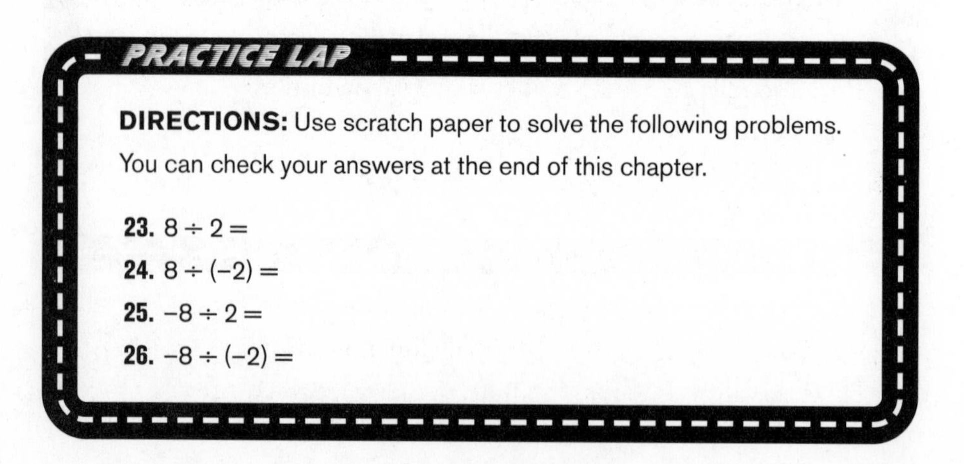

ANSWERS

1. When you consider the set of positive numbers and the set of negative numbers, you are leaving out the number zero. When you add in the set that contains (only) zero, you have three sets that contain all the numbers on the number line. These three sets have no elements in common.
2. The absolute value of –12.4 is 12.4.
3. The absolute value of 12.4 is 12.4.
4. The absolute value of 0 is 0.
5. $-5.43 < 3.17$
6. $-6.43 < -5.43$
7. $-3, -2.7, -2, 2\frac{3}{4}, 4.38, 5$
8. Start at zero. Walk 12 units to the left (get to –12). Then, walk three units to the right. You end up at –9: $-12 + 3 = -9$.
9. Start at zero. Walk 12 units to the left (get to –12), and then walk three more units to the left. You end up at –15: $-12 + (-3) = -15$.
10. Start at zero. Walk 12 units to the right (get to 12), and then walk three units to the left. You end up at 9: $12 - 3 = 12 + (-3) = 9$.
11. The addends, –3.847 and 9.241, have opposite signs, so you should take their absolute values, which are 3.847 and 9.241. Subtract the larger absolute value minus the smaller absolute value:

$$\begin{array}{r} 9.241 \\ -\ 3.847 \\ \hline 5.394 \end{array}$$ (This is the absolute value of the result.)

The addend with the larger absolute value is the positive one, so the sign of your answer is positive: $-3.847 + 9.241 = 5.394$

12. Both addends are positive. This is ordinary addition.

$$\begin{array}{r} 2.104 \\ +\ 3.190 \\ \hline 5.294 \end{array}$$ (here you pad with a zero)

$2.104 + 3.19 = 5.294$

13. This is a subtraction problem, so you change it to add the opposite. $-2.104 - 3.19 = -2.104 + (-3.19)$. Now it is an adding problem. The signs are both the same (namely, negative). So you add the absolute

values and then give the result a negative sign. As in the previous problem, 2.104 + 3.19 = 5.294. The answer is –2.104 – 3.19 = –5.294.

14. The addends have opposite signs, so you subtract the absolute values:

$$\begin{array}{r} 35.14 \\ -12.34 \\ \hline 22.80 \end{array}$$ (This is the absolute value of the result.)

The addend with the larger absolute value, namely –35.14, is negative. So the result is negative. The answer is –35.14 + 12.34 = –22.80.

15. 5 – 3 + 4 – 2 + 10 – 4 = 5 + (–3) + 4 + (–2) + 10 + (–4). Add all the positive numbers: 5 + 4 + 10 = 19. Add all the negative numbers: –3 + (–2) + (–4) = –9. The answer is 19 + (–9) = 10.

16. Start at zero. Walk eight units to the right (get to 8), and then walk five units to the left. You get to 3: 8 – 5 = 8 + (–5) = 3.

17. Start at zero. Walk five units to the right (get to 5), and then walk eight units to the left. Get to –3: 5 – 8 = 5 + (–8) = –3.

18. –3.56 – (–8.62) = –3.56 + (8.62). Now it's an exercise in adding signed numbers. The addends, namely –3.56 and 8.62, have opposite signs, so you subtract the absolute values, which are 3.56 and 8.62:

$$\begin{array}{r} 8.62 \\ -3.56 \\ \hline 5.06 \end{array}$$ (This is the absolute value of the result.)

Take the sign of the addend with the largest absolute value, namely 8.62, which is positive. So, the answer is –3.56 – (–8.62) = 5.06.

19. 12.8 – (–4) = 12.8 + 4. This is now an ordinary addition problem. The answer is 16.8.

20. Negative times positive = negative: (–1) × 17 =–17.

21. (–1) × (–1) × (–1) × (–1) × (–1) = –1. There is an odd number of negatives, so the product is negative.

22. (–1) × (–1) × (–1) × 0 × (–1) × (–1) = 0. The first three factors multiply out to –1. Then, multiplying by 0, you get 0, because 0 times any number is 0. The products of 0 with the next factors are all zero, for the same reason.

23. Positive divided by positive equals positive: 8 ÷ 2 = 4.

24. Positive divided by negative equals negative: 8 ÷ (–2) = –4.

25. Negative divided by positive equals negative: –8 ÷ 2 = –4.

26. Negative divided by negative equals positive: –8 ÷ (–2) = 4.

8 Introducing Algebra

WHAT'S AROUND THE BEND

- Variables
- Like Terms
- The Distributive Law
- Solving Equations

VARIABLES

In algebra, letters, called **variables**, are often used to represent numbers. The a in the equation $a + 3 = 7$ is a variable.

Once you realize that these variables are just numbers in disguise, you'll understand that they must obey all the rules of mathematics, just like the numbers that aren't "disguised" as letters. This can help you figure out what number the variable at hand stands for.

What is the **solution set** of a variable in an equation? It is the set of numbers that can be put in the variable's little box to make the equation true. For example, there is a solution set for the variable a in the equation $a + 3 = 7$. That solution set is the set that contains only the number 4 and is written $\{4\}$.

What is the solution set of the variable x in the equation $x^2 = 9$? This equation can be made true by substituting either 3 or -3 for x. So, the solution set is $\{3,-3\}$.

The Step to Algebra

What is the value, in pennies, of 3 dimes?
3 dimes are worth 30 pennies.

What is the value, in pennies, of 7 dimes?
7 dimes are worth 70 pennies.

Okay, what is the value, in pennies, of d *dimes?*

PACE YOURSELF

THE NEXT TIME you go shopping, take note of the prices on any two items. Use a variable to represent the cost of the first item and a different variable to represent the cost of the second item. Use the variables to write an algebraic expression that will calculate what you spent on the combination of the two items. Evaluate the expression to answer the problem.

What you want to hear is "10 times d," period. Or "10d." Period. You understand that "10d" means the number of pennies that d dimes are worth is ten times the number of dimes. You understand that it is meaningful to say, "d dimes are worth 10d cents," even when you don't know what the value of d is.

When you are asked to *evaluate* an algebraic expression, you substitute a number in place of a variable (letter) and then simplify the expression. Study the following examples.

Example

Evaluate the expression $2b + a$ when $a = 2$ and $b = 4$.

Substitute 2 for the variable a and 4 for the variable b. When the expression is written as $2b$, it means 2 times b.

Multiply 2×4.

$2(4) + \mathbf{2}$
$8 + 2$
$= 10$.

Example

- Evaluate the expression $a^2 + 2b + c$ when $a = 2$, $b = 3$, and $c = 7$.
  Substitute 2 for a, 3 for b, and 7 for c: $a^2 + 2b + c$
  Find the value of 2^2: $(\mathbf{2})^2 + 2(\mathbf{3}) + \mathbf{7}$
  Multiply $2 \cdot 3$: $\mathbf{4} + 2(3) + 7$
  Add the numbers: $4 + \mathbf{6} + 7 = 17$

CAUTION!

WHEN WORKING WITH variables, it is sometimes easy to mistake the variable x for the multiplication sign and vice versa. To avoid this, mathematicians will use the multiplication dot in place of the usual multiplication symbol.

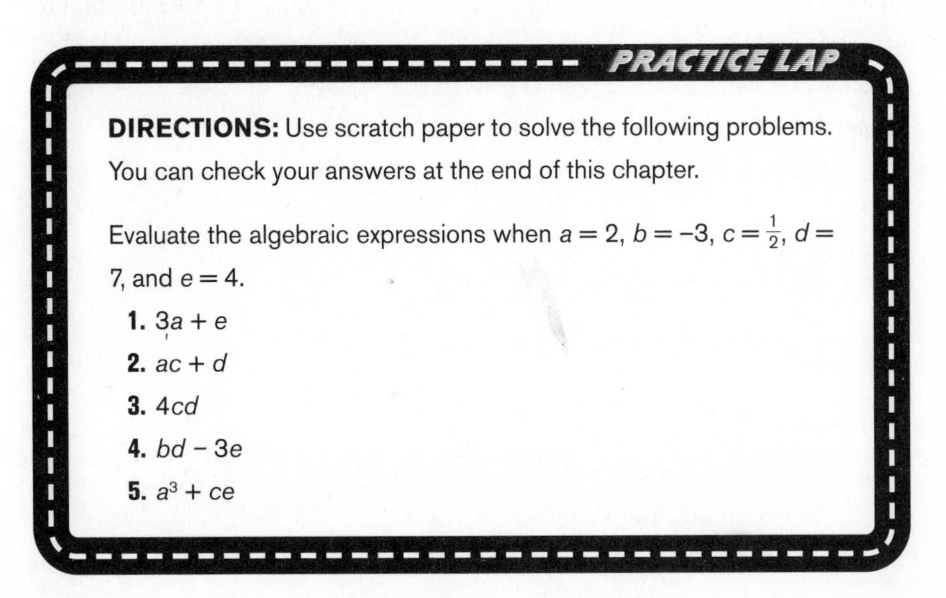

PRACTICE LAP

DIRECTIONS: Use scratch paper to solve the following problems. You can check your answers at the end of this chapter.

Evaluate the algebraic expressions when $a = 2$, $b = -3$, $c = \frac{1}{2}$, $d = 7$, and $e = 4$.

1. $3a + e$
2. $ac + d$
3. $4cd$
4. $bd - 3e$
5. $a^3 + ce$

WHAT ARE LIKE TERMS?

First of all, what are terms? **Terms** are connected by mathematical operators (addition, subtraction, multiplication, or division signs). For example, the expression $a + b$ has two terms and the expression ab has one term. Remember, ab means a times b. The expression ab is one term because it is connected with an understood multiplication sign. The expression $3a + b + 2c$ has three terms. The expression $3ab + 2c$ has two terms.

Secondly, what are like terms and why are they important? **Like terms** have the same variable(s) with the same exponent, such as $3x$ and $10x$. More examples of like terms are:

$3ab$ and $7ab$

$2x^2$ and $8x^2$

$4ab^2$ and $6ab^2$

5 and 9

You can add and subtract like terms. When you add and subtract like terms, you are simplifying an algebraic expression.

How do you add like terms? Simply add the numbers in front of the variables and keep the variables the same. The numbers in front of the variables are called **coefficients**. Therefore, in the expression $6x + 5$, the coefficient is 6.

Example

$2x + 3x + 7x$

Add the numbers in front of the variables: $2 + 3 + 7$
Don't change the variable: $= 12x$

Example

$4xy + 3xy$

Add the numbers in front of the variables: $4 + 3$
Don't change the variables: $= 7xy$

Example

$2x^2y - 5x^2y$

Subtract the numbers in front of the variables: $2 - 5$
Don't change the variables: $= -3x^2y$

Example

$4x + 2y + 9 + 6x + 2$

(Hint: You can add only the like terms, $4x$ and $6x$, and the numbers 9 and 2.)
Add the like terms: $4x + 2y + 9 + 6x + 2$
$= 10x + 2y + 11$

Using the Distributive Property to Combine Like Terms

What do you do with a problem like this: $2(x + y) + 3(x + 2y)$? According to the order of operations, you would have to do the grouping symbols first. However, you know you can't add x to y because they are not like terms. What you need to do is use the **distributive property**. The distributive property tells you to multiply the number and/or variable(s) outside the parentheses by every term inside the parentheses.

FUEL FOR THOUGHT

IN ORDER TO simplify an expression that contains several different operations (addition, subtraction, multiplication, division, etc.), it is important to do them in the right order. This is called the **order of operations.**

For example, suppose you were asked to simplify the expression $3 + 4 \cdot 2 + 5$. At first glance, you might think it is easy: $3 + 4 = 7$ and $2 + 5 = 7$, then $7 \cdot 7 = 49$. Another person might say $3 + 4 = 7$ and $7 \cdot 2 = 14$ and $14 + 5 = 19$. Actually, both of these answers are wrong!

To eliminate the possibility of getting several answers for the same problem, you must follow the order of operations:

1. Perform the operations inside grouping symbols such as ( ), { }, and [ ]. The division bar can also act as a grouping symbol. The division bar or fraction bar tells you to do the steps in the numerator and the denominator before you divide.
2. Evaluate exponents (powers) such as 3^2.
3. Do all multiplication and division *in order from left to right.*
4. Do all addition and subtraction *in order from left to right.*

In order to remember that the order of operations is **P**arentheses, **E**xponents, **M**ultiplication & **D**ivision, then **A**ddition & **S**ubtraction, you can use the sentence: "**P**lease **e**xcuse **m**y **d**ear **A**unt **S**ally."

You would work the problem $2(x + y) + 3(x + 2y)$ like this:

Multiply 2 times x and 2 times y. Then multiply 3 times x and 3 times $2y$. If there is no number in front of the variable, it is understood to be 1, so 2 times x means 2 times $1x$. To multiply, you multiply the numbers and the variable stays the same. When you multiply 3 times $2y$, you multiply 3 times 2 and the variable, y, stays the same, so you would get $6y$. After you have multiplied, you can then combine like terms.

Example

Multiply $2(x + y)$ and $3(x + 2y)$.

Combine like terms: $2x + 2y + 3x + 6y$

$= 5x + 8y$

INSIDE TRACK

IF THERE IS no number in front of a variable, it is understood to be 1.

Here are two more examples using the distributive property.

Example

$2(x + y) + 3(x - y)$

Multiply 2 times x and 2 times y. Then, multiply 3 times x and 3 times $(-y)$. When you multiply 3 times $(-y)$, this is the same as $3(-1y)$. The 1 is understood to be in front of the y even though you don't see it. In this example, you can see how the parentheses are used to indicate multiplication: $2(x + y) + 3(x - y)$

Use the distributive property: $2x + 2y + 3x - 3y$

Combine like terms.

$= 5x - y$

Example

$2(2x + y) - 3(x + 2y)$

Use the distributive property to get rid of the parentheses. The subtraction sign in front of the 3 is the same as multiplying $(-3)(x)$ and $(-3)(2y)$.

Use the distributive property: $2(2x + y) - 3(x + 2y)$

Combine like terms: $4x + 2y - 3x - 6y$

$= x - 4y$

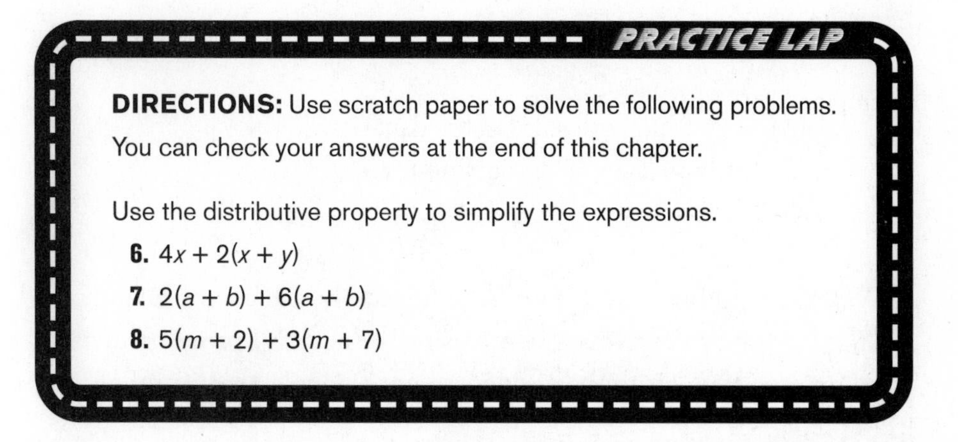

PRACTICE LAP

DIRECTIONS: Use scratch paper to solve the following problems. You can check your answers at the end of this chapter.

Use the distributive property to simplify the expressions.

6. $4x + 2(x + y)$

7. $2(a + b) + 6(a + b)$

8. $5(m + 2) + 3(m + 7)$

SOLVING EQUATIONS

An equation is a mathematical tool that helps people solve many real-life problems. What is an equation? An **equation** is two equal expressions. These expressions could be numbers such as $6 = 5 + 1$ or variable expressions such as $D = rt$. What does it mean to solve an equation? When you find the value of the variable, you have solved the equation. For example, you have solved the equation $2x = 10$ when you know the value of x.

Here is the basic rule for solving equations: When you do something to one side of an equation, you must do the same thing to the other side of the equation. You'll know that you have solved an equation when the variable is alone (isolated) on one side of the equation and the variable is positive. In the example $-x = 5$, 5 is not the answer because the x variable is negative.

You can solve equations using addition and subtraction by getting the variable on one side of the equal sign by itself. Think about how you would solve this equation: $x + 4 = 10$. Your goal is to get the variable x on a side by itself. How do you do that?

To get x by itself, you need to get rid of the 4. How? If you subtract 4 from 4, the result is 0, and you have eliminated the 4 to get x on a side by itself. However, if you subtract 4 from the left side of the equation, then you must do the same to the right side of the equation. You would solve the equation this way:

Example

$x + 4 = 10$

Subtract 4 from both sides of the equation: $x + 4 - 4 = 10 - 4$
Simplify both sides of the equation: $x + 0 = 6$
Add 0 to x: $x = 6$

When you add zero to a number, the number does not change, so $x + 0 = x$. When you do this, you are using the **additive property of zero**, which states that a number added to zero equals that number. For example, $5 + 0 = 5$.

Let's look at another equation: $x - 5 = 9$. What do you need to do to get the variable on a side by itself? You need to get rid of the 5. The equation says to subtract 5, so what undoes subtraction? If you said addition, you are right! In this problem, you need to add 5 to both sides of the equation to get rid of the 5.

Example

$x - 5 = 9$

Add 5 to both sides of the equation: $x - 5 + 5 = 9 + 5$
Simplify both sides of the equation: $x + 0 = 14$
Add 0 to x: $x = 14$

Example

$a + 6 = 7$

Subtract 6 from both sides of the equation: $a + 6 - 6 = 7 - 6$
Simplify both sides of the equation: $a + 0 = 1$
Add 0 to a: $a = 1$

Example

$y - 11 = 8$

Add 11 to both sides of the equation: $y - 11 + 11 = 8 + 11$
Simplify both sides of the equation: $y + 0 = 19$
Add 0 to y: $y = 19$

Example

$-r + 9 = 13$

Subtract 9 from both sides of the equation: $-r + 9 - 9 = 13 - 9$
Simplify both sides of the equation: $-r + 0 = 4$
Add 0 to $-r$: $-r = 4$

Are you finished? No! The value of the variable must be positive, and r is negative. You need to make the variable positive. You can do that by multiplying both sides of the equation by (-1).

$(-1)(-r) = 4(-1)$

$r = -4$

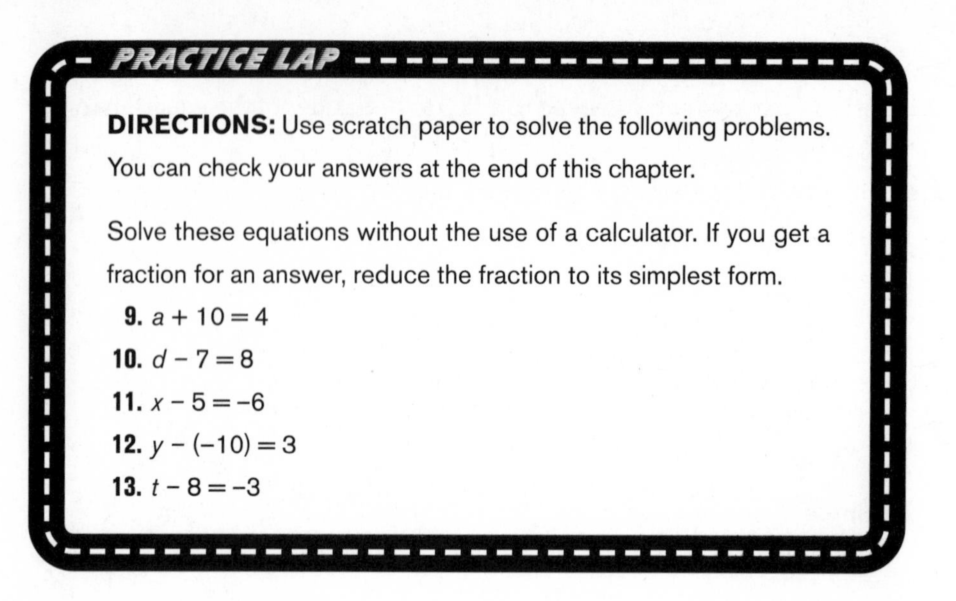

PRACTICE LAP

DIRECTIONS: Use scratch paper to solve the following problems. You can check your answers at the end of this chapter.

Solve these equations without the use of a calculator. If you get a fraction for an answer, reduce the fraction to its simplest form.

9. $a + 10 = 4$

10. $d - 7 = 8$

11. $x - 5 = -6$

12. $y - (-10) = 3$

13. $t - 8 = -3$

Checking Your Answers

You can use the answer key at the end of this lesson to check your answers. However, in the world of work, there is no answer key to tell you if you got the correct answer to an equation. There is a way to check your answer. When you replace the variable with your answer, you should get a true statement. If you get a false statement, then you do not have the right answer. To check practice problem 9 in this lesson, you would do the following steps.

***Check for Problem 9:* $a + 10 = 4$**

The answer you got for problem was -6, so replace the variable, which is a, with -6. This is called *substitution.*

You are substituting the variable with the number -6.
Substitute -6 in place of the variable, a: $-6 + 10 = 4$
Simplify the left side of the equation: $4 = 4$

Your result, $4 = 4$, is a true statement, which tells you that you solved the equation correctly.

***Check for Problem 10:* $d - 7 = 8$**

Use the substitution method again.

Substitute 15, your answer, in place of the variable: $15 - 7 = 8$
Simplify the left side of the equation: $8 = 8$

Because the result is a true statement, you know you solved the equation correctly.

Solving Equations Using Multiplication and Division

In the equation $x + 10 = 2$, to get rid of the 10, you would subtract 10, which is the opposite of adding 10. In the equation $x - 5 = 6$, to get rid of the 5, you would add 5, which is the opposite of subtracting 5. So in the equation $5x = 10$, how do you think you would get rid of the 5? What is the opposite of multiplying by 5? Yes, the opposite of multiplying by 5 is dividing by 5. You would solve this equation using division.

$$5x = 10$$

$$\frac{5x}{5} = \frac{10}{5}$$

$$x = 2$$

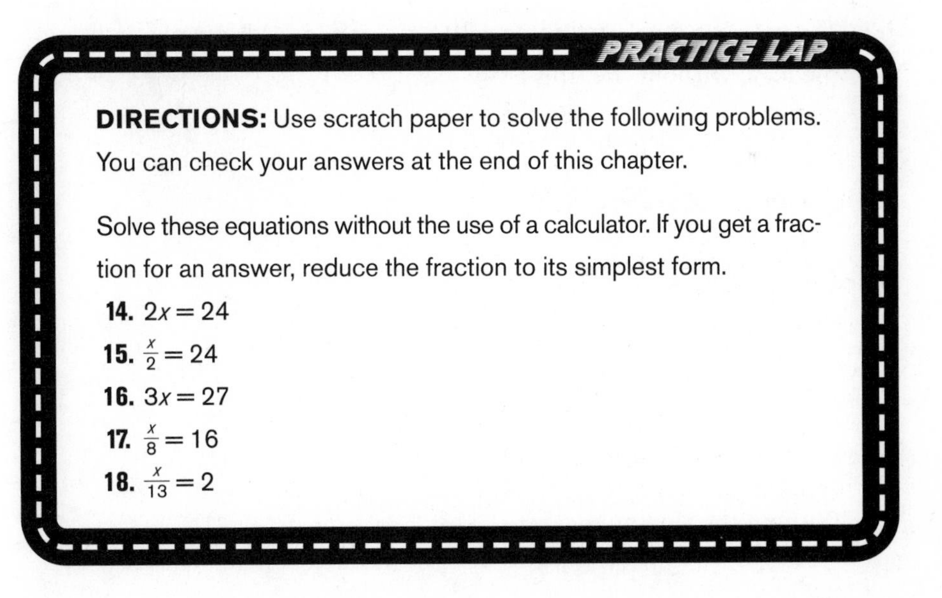

PRACTICE LAP

DIRECTIONS: Use scratch paper to solve the following problems. You can check your answers at the end of this chapter.

Solve these equations without the use of a calculator. If you get a fraction for an answer, reduce the fraction to its simplest form.

14. $2x = 24$

15. $\frac{x}{2} = 24$

16. $3x = 27$

17. $\frac{x}{8} = 16$

18. $\frac{x}{13} = 2$

Setting Up Equations for Word Problems

Equations can be used to solve real-life problems. In an equation, the variable often represents the answer to a real-life problem. For example, suppose you know that you can earn twice as much money this summer as you did last summer. You made $1,200 last summer. How much will you earn this summer? You can use the variable x to represent the answer to the problem. You want to know how much you can earn this summer.

Let x = how much you can earn this summer

$1,200 = amount earned last summer

$x = 2 \cdot 1{,}200$

$x = \$2{,}400$

You might be wondering why you should use algebra to solve a problem when the answer can be found using arithmetic.

Good question! If you practice using equations for simple problems, then you will find it easier to write equations for problems that can't be solved using arithmetic.

Example

The cost of a meal, including a 20% tip, is $21.60. How much was the meal without the tip?

Let x = cost of the meal without the tip.

The tip is 20% of x, or $0.2x$.

Together, $x + 0.2x = 21.60$.

$1.2x = 21.6$

$\frac{1.2x}{1.2} = \frac{21.6}{1.2}$

$x = 18$

Without the tip, the meal cost $18.

Example

There are twice as many girls in your algebra class as there are boys. If there are 18 girls in the class, how many boys are in the class?

Let x = the number of boys in the class.

Then 2 times the number of boys is equal to the number of girls. You can represent that in the equation $2x = 18$.

$x = 9$; therefore, there are 9 boys in the class.

Example

You are going to be working for Success Corporation. You got a signing bonus of \$2,500, and you will be paid \$17.50 an hour. If you are paid monthly, how much will your first paycheck be? Be sure to include your signing bonus, and assume that you have a 40-hour workweek and there are 4 weeks in this month.

Let x = first monthly paycheck

Then $x = 4 \cdot 40 \cdot \$17.50 + \$2{,}500$

$x = \$5{,}300$

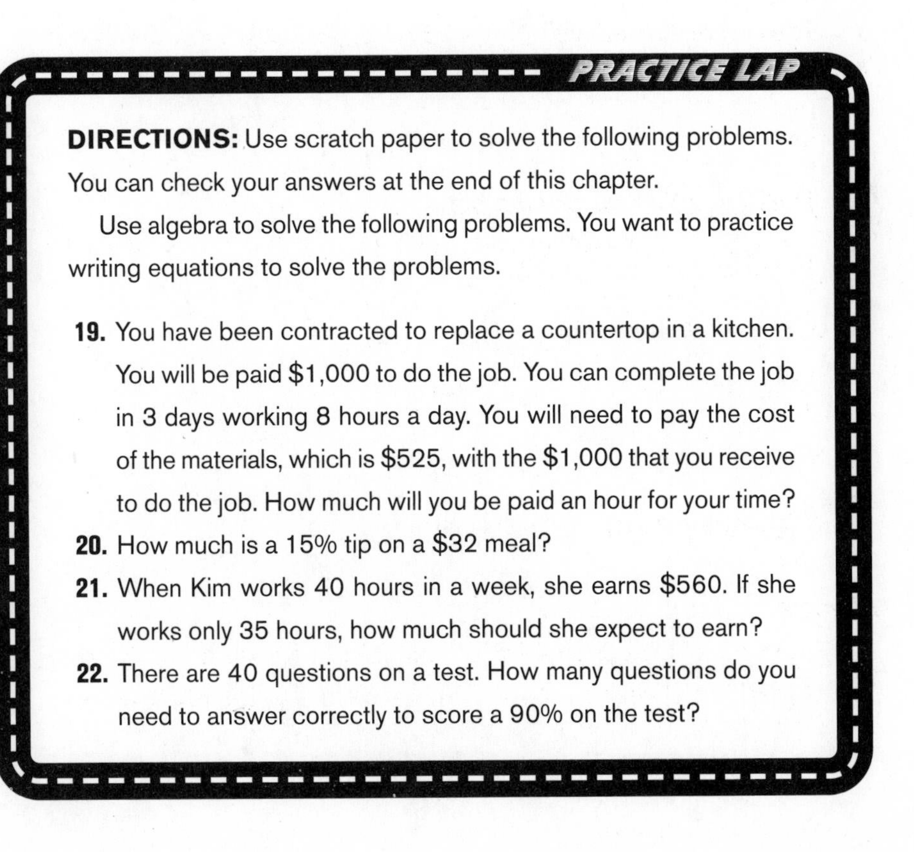

PRACTICE LAP

DIRECTIONS: Use scratch paper to solve the following problems. You can check your answers at the end of this chapter.

Use algebra to solve the following problems. You want to practice writing equations to solve the problems.

19. You have been contracted to replace a countertop in a kitchen. You will be paid \$1,000 to do the job. You can complete the job in 3 days working 8 hours a day. You will need to pay the cost of the materials, which is \$525, with the \$1,000 that you receive to do the job. How much will you be paid an hour for your time?

20. How much is a 15% tip on a \$32 meal?

21. When Kim works 40 hours in a week, she earns \$560. If she works only 35 hours, how much should she expect to earn?

22. There are 40 questions on a test. How many questions do you need to answer correctly to score a 90% on the test?

ANSWERS

1. 10
2. 8
3. 14
4. –33
5. 10
6. $6x + 2y$
7. $8a + 8b$
8. $8m + 31$
9. $a = -6$
10. $d = 15$
11. $x = -1$
12. $y = -7$
13. $t = 5$
14. 12
15. 48
16. 9
17. 128
18. 26
19. $x = \frac{(\$1{,}000 - \$525)}{24 \text{ hours}}$, \$19.79 per hour
20. $x = 0.15 \cdot \$32$, $x = \$4.80$
21. $40 \cdot \frac{x}{35} = \$560$, $x = \$490$
22. $x = .90 \cdot 40$, 36 questions

9 Geometry and Measurement Conversions

WHAT'S AROUND THE BEND

- Polygons
- Classifying Triangles
- Classifying Quadrilaterals
- Perimeter
- Area
- Circles
- Pythagorean Theorem
- Volume
- Measurement Conversions

POLYGONS

Plane figures are two-dimensional objects that reside on a plane. You can think of a plane like a sheet of paper that extends forever in all directions. Special figures are called **polygons**, several of which are defined here.

FUEL FOR THOUGHT

SOME IMPORTANT WORDS to know:

A **polygon** is a closed plane figure made up of line segments.
A **triangle** is a polygon with three sides.
A **quadrilateral** is a polygon with four sides.
A **pentagon** is a polygon with five sides.
A **hexagon** is a polygon with six sides.
An **octagon** is a polygon with eight sides.

Polygons are made up of angles and line segments called **sides**. Each angle is made up of two sides, and the point at which they meet is called the **vertex**.

TRIANGLES

Triangles are three-sided polygons. Triangles are classified, or grouped, in two different ways. One classification distinguishes by the sides, and another by the angles. For a triangle, you can have all three sides **congruent** (equal measure), two sides congruent, or no sides congruent.

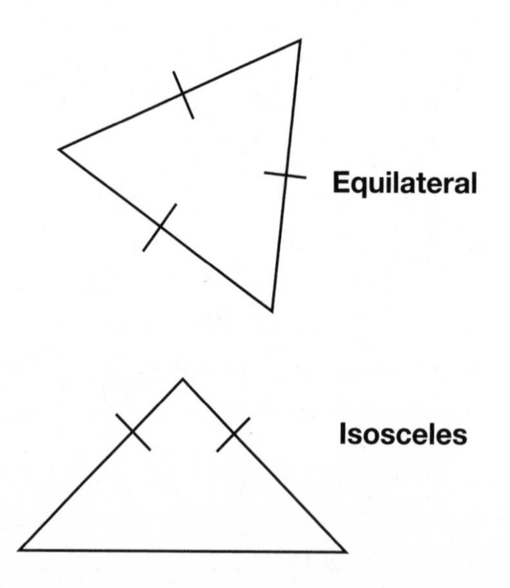

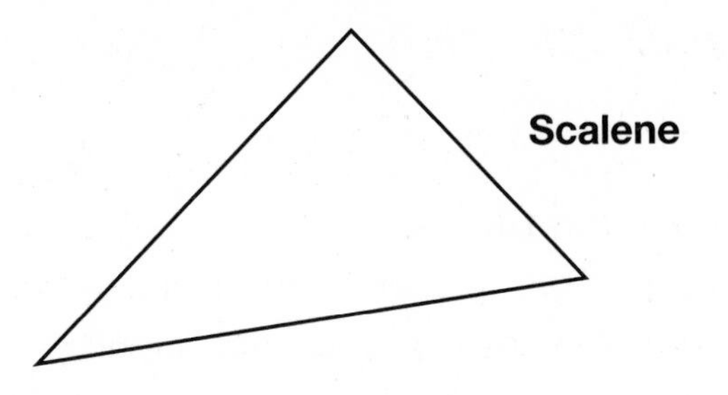

In triangle figures, the little box drawn inside an angle stands for a right angle (90°). Here is the classification for triangles when grouped by angle:

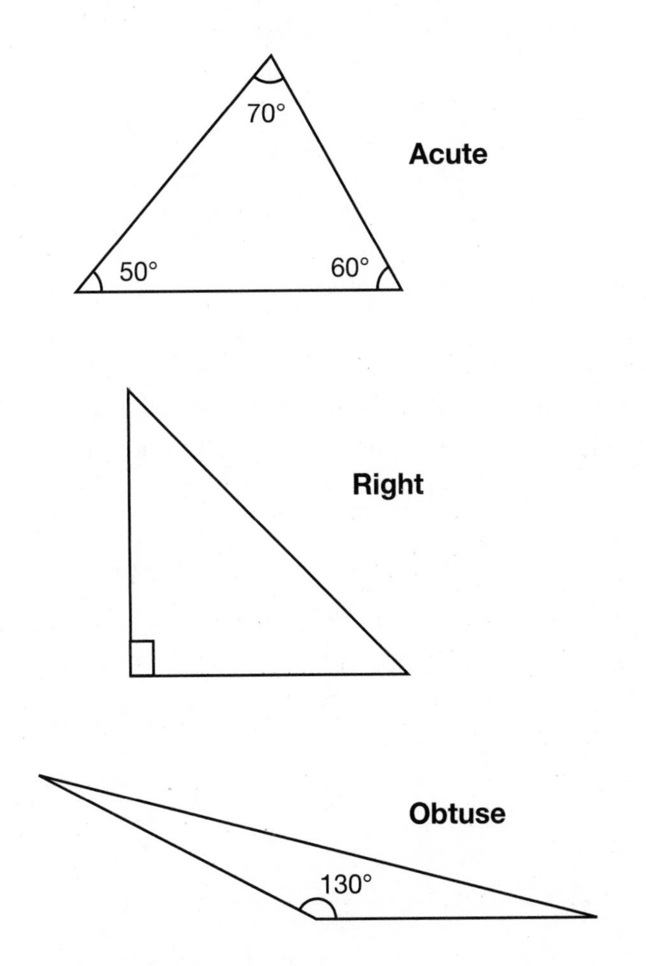

Note that even though right triangles and obtuse triangles each have two acute angles, their classification is not affected by these angles. Acute triangles have all THREE acute angles.

QUADRILATERALS

Four-sided polygons are called **quadrilaterals**, and like triangles, there are classifications for quadrilaterals.

A quadrilateral with one pair of parallel sides (bases) is called a **trapezoid**.

In an **isosceles trapezoid**, the sides that are not bases are congruent. Because the parallel bases are not the same length in a trapezoid, we call these bases b_1 and b_2.

A quadrilateral with two pairs of parallel sides is called a **parallelogram**. The two sets of opposite sides that are parallel are congruent in a parallelogram, as shown in the figure:

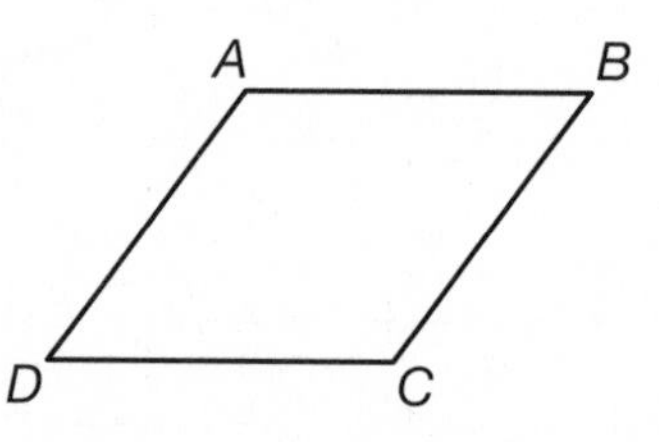

Parallelograms are broken down into further subgroups.

A **rectangle** is a parallelogram with four right angles.

A **rhombus** is a parallelogram with four congruent sides.

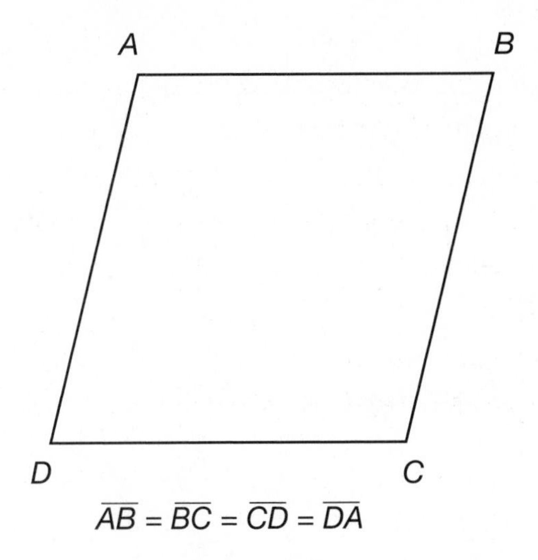

A **square** is a parallelogram with both four right angles and four congruent sides.

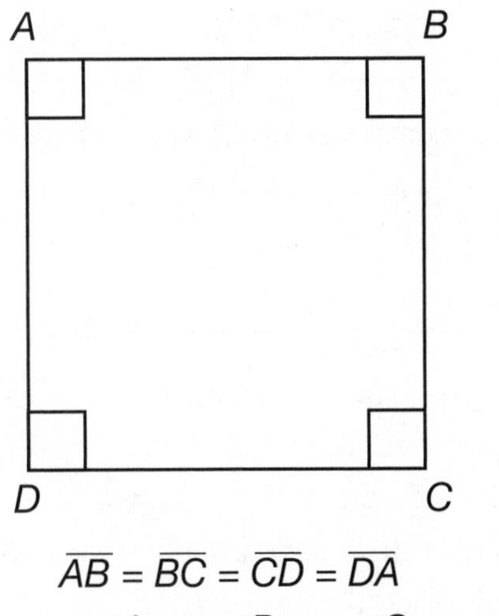

$\overline{AB} = \overline{BC} = \overline{CD} = \overline{DA}$

$m\angle A = m\angle B = m\angle C = m\angle D$

PERIMETER

Perimeter is the measure AROUND a polygon. Perimeter is an addition concept; it is a linear, one-dimensional measurement.

To find the perimeter of a polygon, add up all of the lengths of the sides of the figure. Be sure to name the units.

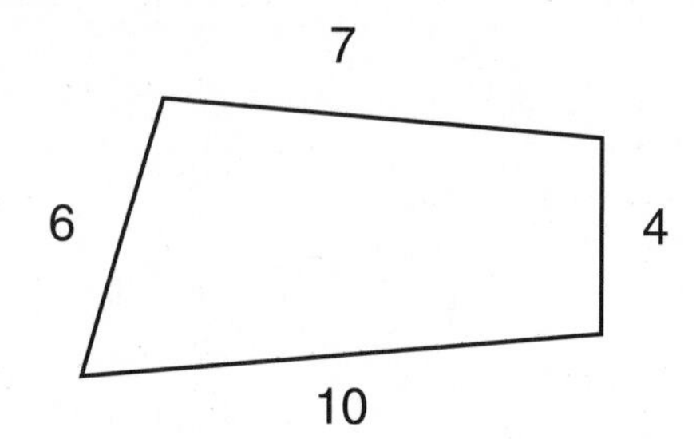

Perimeter = 6 + 7 + 4 + 10 = 27

INSIDE TRACK

FOR A RECTANGLE, perimeter can be found using the formula $P = 2l + 2w$, where P is the perimeter, l is the length, and w is the width.

For a square, or any rhombus, the perimeter can be found by $P = 4s$, where P is the perimeter and s is the length of one of the sides.

Example

Find the perimeter of a square whose side is 5 centimeters.

You know that each side of a square is equal. So, if you know that one side of a square is 5 centimeters, then you know that each side of the square is 5 centimeters. To calculate the perimeter, add up the length of all four sides.

5 + 5 + 5 + 5 = 20

So, the perimeter of a square whose side measures 5 cm is 20 centimeters.

Be alert when working with geometry problems to make sure that the units are consistent. If they are different, a conversion must be made before calculating perimeter or area.

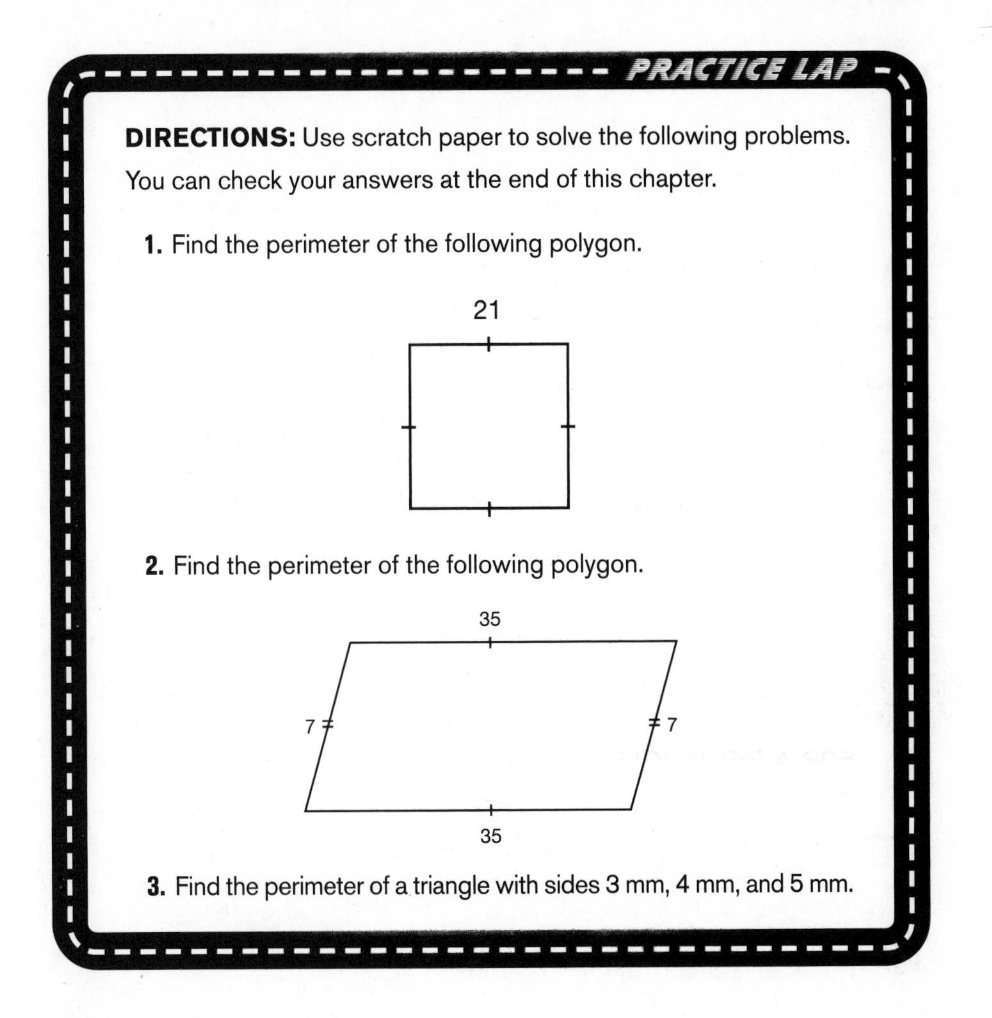

PRACTICE LAP

DIRECTIONS: Use scratch paper to solve the following problems. You can check your answers at the end of this chapter.

1. Find the perimeter of the following polygon.

2. Find the perimeter of the following polygon.

3. Find the perimeter of a triangle with sides 3 mm, 4 mm, and 5 mm.

AREA

Area is a measure of how many square units it takes to COVER a closed figure. Area is measured in square units. Area is a multiplication concept, where two measures are multiplied together. You can also think of units being multiplied together: cm $\times$ cm = cm^2, or the words *centimeters squared.* There are formulas to use for the area of common polygons:

Common Polygon Formulas

A stands for area, b stands for base, h stands for height (which is perpendicular to the base), and b_1 and b_2 are the parallel sides of a trapezoid.

Figure	Area calculation	Area formula
Square	**side × side** *or* **base × height**	$A + s^2$ $A = bh$
Rectangle	**base × height**	$A = bh$
Parallelogram	**base × height**	$A = bh$
Triangle	$\frac{1}{2}$ **× base × height**	$A = \frac{1}{2}bh$
Trapezoid	$\frac{1}{2}$ **× base**$_1$ **× height ×** $\frac{1}{2}$**base**$_2$ **× height**	$A = \frac{1}{2}h(b_1 + b_2)$

Notice that to calculate the area of a figure, you need two measurements. Multiplying the units of each measurement together gives you *square units.* That's why area is measured in square units. (Perimeter was measured in *linear* units.) Be sure to include square units in your answer.

Examples of square units are

square miles or mi^2

square meters or m^2

square centimeters or cm^2

square feet or $ft.^2$

square inches or $in.^2$

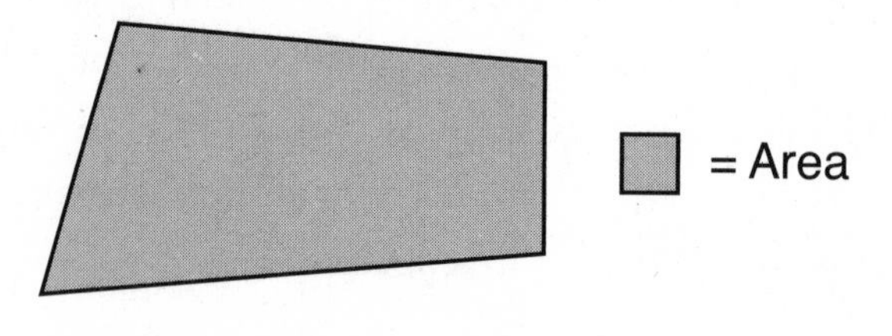

CIRCLES

Circles are another common plane figure.

FUEL FOR THOUGHT

SOME IMPORTANT WORDS to know:

A **circle** is the set of all points equidistant from one given point, called the **center**. The center point defines the circle but is not on the **circle**.

A **diameter** of a circle is a line segment that passes through the center of the circle and whose endpoints are on the circle. The diameter is twice the radius of the circle; $d = 2r$.

A **radius** of a circle is any line segment whose one endpoint is at the center of the circle and whose other endpoint is on the circle. The radius is one-half the length of the diameter; $r = \frac{1}{2}d$.

The **circumference** of a circle is the distance AROUND the circle (the perimeter).

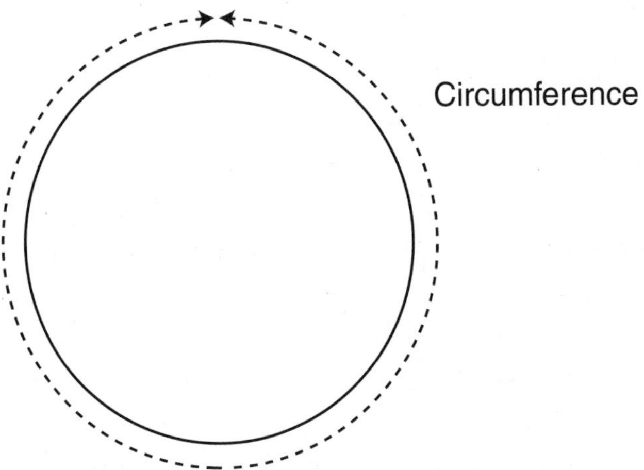

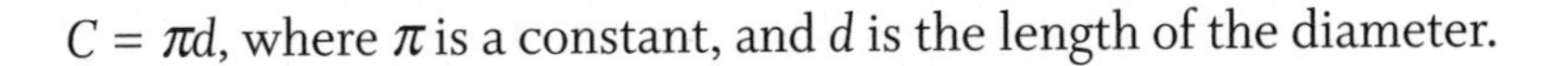

$C = \pi d$, where π is a constant, and d is the length of the diameter.

OR

$C = 2\pi r$, where r is the length of the radius.

Example

Find the circumference of a circle with a diameter of 5 inches.

Because you know the diameter, use the formula that includes the diameter: $C = \pi d$.

$C = \pi(5)$

$= (3.14)(5)$

$= 15.7$

The final answer is 15.7 inches.

INSIDE TRACK

IF YOU'RE NOT allowed to use a calculator on a test, you can estimate the circumference of a circle by substituting 3 for π.

Area of a circle is the number of square units it takes to cover the circle. $A = \pi r^2$, where π is a constant and r is the radius.

π, called "pi," is a special ratio that is a constant value of approximately 3.14. Pi compares circumference to diameter in the following ratio: $\pi = \frac{C}{d}$. It is the same value for every circle. Often, in math tests, answers will be given in terms of π, such as 136π square units. If answers are not given in terms of π, use the π key on your calculator unless otherwise instructed. Sometimes, a problem will direct you to use either $\pi = 3.14$, or $\pi = \frac{22}{7}$, which are approximations for pi.

Using the formulas from the previous paragraph, you can calculate the circumference and area of circles. Take care and check if the problem gives the radius or diameter. If the problem asks for the area of the circle, for example, and gives the length of the diameter, you must first calculate the length of the radius. The radius can be found by dividing the diameter by two. Just like for all area calculations, the units will be square units. The units for circumference will be linear (single) units.

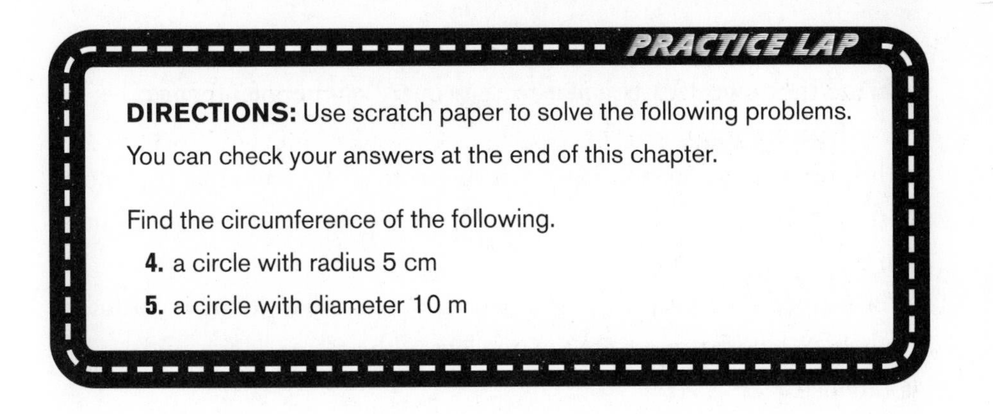

PRACTICE LAP

DIRECTIONS: Use scratch paper to solve the following problems. You can check your answers at the end of this chapter.

Find the circumference of the following.

4. a circle with radius 5 cm

5. a circle with diameter 10 m

RIGHT TRIANGLE MEASUREMENTS: PYTHAGOREAN THEOREM

Right triangles are special triangles used for measuring. In a right triangle, the base and one side are perpendicular; that is, they form a 90° angle.

FUEL FOR THOUGHT

SOME IMPORTANT WORDS to know:

The **hypotenuse** of a right triangle is the side of the right triangle that is opposite the right angle.

The **legs** of a right triangle are the two sides of the right triangle that make up the right angle.

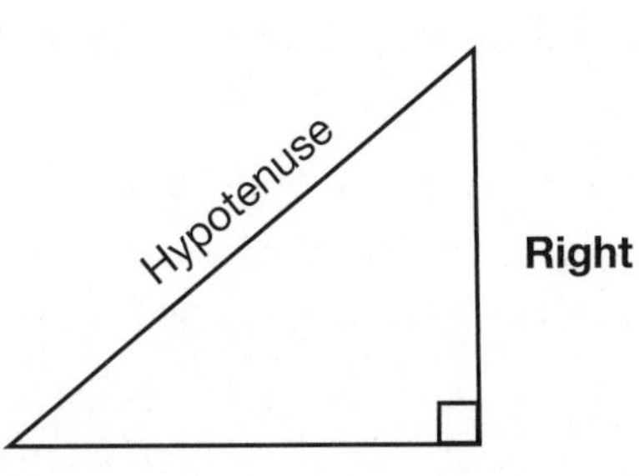

In right triangles, there is a special relationship between the hypotenuse and the legs of the triangle. This relationship is always true, and it is known as the Pythagorean theorem.

The **Pythagorean theorem** states that in all right triangles, the sum of the squares of the two legs is equal to the square of the hypotenuse: $\text{leg}^2 + \text{leg}^2 = \text{hypotenuse}^2$.

The **converse of the Pythagorean theorem** is also true: In a triangle, if the sum of the squares of the legs is equal to the square of the hypotenuse, then the triangle is a right triangle.

You can remember the Pythagorean theorem as the well-known formula: $a^2 + b^2 = c^2$, where a and b are the two legs of the right triangle and c is the hypotenuse.

Special note: Be careful! There is nothing special about the letters a, b, and c. A test question could be "tricky" and could call one of the legs "c."

INSIDE TRACK

THERE ARE THREE sets of Pythagorean triples that appear over and over again in math test problems. Knowing these three common triples will save you valuable time in working problems of this type.

	a	***b***	***c***
One set is:	3	4	5
and multiples thereof:	6	8	10
	9	12	15
	12	16	20
Another set is:	5	12	13
and multiples thereof:	10	24	26
	15	36	39
The third set is:	8	15	17
	16	30	34

Memorize these sets: {3, 4, 5}, {5, 12, 13}, and {8, 15, 17}. If a right triangle problem is given and two of the three numbers in one set (or multiples of the two numbers) appear, you can avoid all the substituting and calculating and save precious test time.

VOLUME

Volume is a measure of how many cubic units it takes to FILL a solid figure. Volume is measured in cubic units. You may recall the term *cube* from Chapter 7. Volume, like area, is a multiplication concept, where three measures are multiplied together. Unlike area, volume is three-dimensional. The units can also be thought of as multiplied together: cm × cm × cm = cm^3, or the words *centimeters cubed.*

Volume is the area of the base of the solid figure, multiplied by the height of the figure. This can be expressed as $V = Bh$, where V is the volume, B is the area of the base, and h is the height.

Volume of a rectangular prism $V = lwh$

in this case, $B = lw$, where l is the length and w is the width.

Volume of a cylinder $V = \pi r^2 h$

in this case, $B = \pi r^2$, where π is the constant, and r is the radius of the cylindrical base.

Example

Given the volume of a rectangular prism to be 362.1 mm^3, find the height if the length is 7.1 mm and the width is 5 mm.

Use the formula for the volume of a rectangular prism:

$V = lwh$	Substitute in the given information.
$362.1 = 7.1 \times 5 \times h$	Multiply 7.1 times 5.
$362.1 = 35.5 \times h$	Divide 362.1 by 35.5 and include units.
10.2 mm = h	The units are linear (single) because this is a height measurement.

Example

Find the volume of a cylinder whose base diameter is 12.4 inches and has a height of 15 inches.

First, recognize that while diameter is given, the radius is needed to calculate volume. Use the formula to find the radius:

$r = \frac{1}{2}d$	Substitute in the given value for diameter.
$r = \frac{1}{2} \times 12.4$	Multiply one-half times 12.4.
$r = 6.2$ in.	Now use the formula for the volume of a cylinder.
$V = \pi r^2 h$	Substitute in the given information.
$V = \pi \times 6.2 \times 6.2 \times 15$	Multiply all the number terms together on the right.
$V = 576.6\pi$	Include the cubic units.
$V = 576.6\pi \text{ in.}^3$	

MEASUREMENT CONVERSIONS

When working with measurements, you often have to convert units before performing other calculations. There are two methods of converting measurements. One uses proportions and the other uses a scientific method called **dimensional analysis**.

Before describing the methods of conversion, here are some common conversions that are used:

Conversion Factors

1 foot = 12 inches	1 cup = 8 ounces
3 feet = 1 yard	1 pint = 2 cups
1 mile = 5,280 feet	1 quart = 2 pints
1 minute = 60 seconds	1 gallon = 16 cups
1 hour = 60 minutes	1 gallon = 4 quarts
1 inch = 2.54 centimeters	

Metric Conversions

1 meter = 1,000 millimeters

1 meter = 100 centimeters

1 meter = 10 decimeters

1,000 meters = 1 kilometer

The metric prefixes and their meanings are:

Prefix	Meaning	Example
kilo	1,000 times	1 kilometer is 1,000 meters.
hecto	100 times	1 hectogram is 100 grams.
deca	10 times	1 decaliter is 10 liters.
deci	$\frac{1}{10}$ times	1 decigram is $\frac{1}{10}$ of a gram.
centi	$\frac{1}{100}$ times	1 centimeter is $\frac{1}{100}$ of a meter.
milli	$\frac{1}{1{,}000}$ times	1 milliliter is $\frac{1}{1{,}000}$ of a liter.

The prefixes can be used with any of the following: *meters* measure length, *liters* measure volume, and *grams* measure mass.

To use the proportion method to convert units, set up a proportion. (For information on proportions, see Chapter 10.) Keep the units consistent on both sides of the proportion. For example, if you want to convert 50.8 centimeters to inches, set up a proportion, such as $\frac{\text{inch}}{\text{centimeter}} = \frac{\text{inch}}{\text{centimeter}}$, and substitute in the given values on one side, the conversion factor on the other: $\frac{1}{2.54} = \frac{n}{50.8}$. Cross multiply to get $2.54 \times n = 1 \times 50.8$. Now, divide 50.8 by 2.54 to get $n = 20$ inches.

The dimensional analysis method involves multiplying a series of fractions that are all equal to a value of one, with unwanted units alternately on top and bottom. This way, the units "cancel" and what is left is the needed unit. Again, to convert 50.8 centimeters to inches, set up the analysis: $\frac{1}{2.54}$ centimeters $= 50.8\,\frac{\text{centimeters}}{n}$. The setup for this analysis is the opposite of using a proportion; notice that in this setup, the centimeter units are in opposite positions in the fraction. The centimeter units, which we do not want, will cancel, and all that remains is the needed inch unit.

$$\frac{1 \text{ in.}}{2.54 \cancel{\text{cm}}} \times \frac{50.8 \cancel{\text{cm}}}{n}$$

After canceling the centimeter units and multiplying the fractions straight across, the problem becomes $\frac{50.8 \text{ inches}}{2.54} = 20$ inches.

You may be wondering, "Why bother with dimensional analysis at all?" There is a very good reason to know this method. Scientists prefer the method because it allows you to do several conversions at once. Study the next example. Both methods of converting units will be used.

Example

A toy car travels a distance of 7,620 cm in four minutes. What is the speed in inches per second?

First, let's use the proportion method. To use this method, centimeters must be converted into inches in one step and then minutes must be converted to seconds in another step. Set up the proportion for centimeters to inches by using $\frac{\text{inch}}{\text{centimeter}} = \frac{\text{inch}}{\text{centimeter}}$, which is $\frac{1}{2.54} = \frac{n}{7{,}620}$. Cross multiply to yield $2.54 \times n = 1 \times 7{,}620$. Divide 7,620 by 2.54 to get 3,000 inches. Now, use the proportion $\frac{\text{seconds}}{\text{minutes}} = \frac{\text{seconds}}{\text{minutes}}$ to convert minutes to seconds. The proportion becomes $\frac{60}{1} = \frac{n}{4}$. Cross multiply to get $1 \times n = 60 \times 4$. Sixty times four is 240, so $n = 240$ seconds. Finally, calculate inches per second, which means inches per one second: $\frac{3{,}000}{240} = 12.5$ inches per second.

Now, solve the problem using the dimensional analysis method. We want the answer to be in inches per second. Set up the fractions with inches on the top and seconds on the bottom, so that the centimeter and minute units cancel.

$$\frac{1 \text{ inch}}{2.54 \text{ centimeters}} \times \frac{7{,}620 \text{ centimeters}}{4 \text{ minutes}} \times \frac{1 \text{ minute}}{60 \text{ seconds}}$$

Now, cancel out units:

$$\frac{1 \text{ in.}}{2.54 \cancel{\text{cm}}} \times \frac{7{,}620 \cancel{\text{cm}}}{4 \cancel{\text{min.}}} \times \frac{1 \cancel{\text{min.}}}{60 \text{ seconds}}$$

Finally, multiply straight across and leave the units that did not cancel:

$$\frac{7{,}620 \text{ inches}}{2.54} \times 4 \times 60 \text{ seconds} = \frac{7{,}620}{609.6}$$

Divide 7,620 by 609.6 to get 12.5 inches per second. As you can see from this example, dimensional analysis is an efficient way to convert measurement units when there are several conversions to be made.

ANSWERS

1. You could have added all the sides: $21 + 21 + 21 + 21 = 84$. Or, you could have multiplied 21×4, because this polygon is a square.
2. You could have added all the sides: $35 + 7 + 35 + 7 = 84$. Or, you could have used the formula $P = 2l + 2w$, because this polygon is a rectangle.
3. Add up the lengths of the three sides: $3 + 4 + 5 = 12$. So, your final answer is 12 mm.
4. You are given the radius, so use $C = 2\pi r$. Plug in the radius and pi and multiply: $(2)(3.14)(5) = 31.4$. So, your final answer is 31.4 cm.
5. You are given the diameter, so use $C = \pi d$. Plug in the diameter and pi and multiply: $(3.14)(10) = 31.4$. So, your final answer is 31.4 cm.

10 Ratios, Proportions, and Measures of Central Tendency

WHAT'S AROUND THE BEND

- Ratios
- What's a Proportion?
- Working with Proportions
- Means
- Medians
- Modes
- Range

Ratios and proportions are about comparing numbers. That's something you do all the time in math and in other areas.

Two ways of comparing 30 and 10 are to ask (1) 30 is 10 *plus* how many, and (2) 30 is 10 *times* what number.

Ratios and proportions are about the latter kind of comparison: What do you have to multiply 10 by in order to get 30?

In this case, you have to multiply 10 by 3 to get 30. That means that the ratio of 30 to 10 is 3 to 1.

What is the ratio of a to b? That question means, a is how many times bigger or smaller than b?

GETTING TO KNOW RATIOS

A ratio, or a rate, is a comparison between two quantities. A **ratio** is a comparison of two numbers. For example, let's say that there are 3 men for every 5 women in a particular club. That means that the ratio of men to women is 3 to 5. It doesn't necessarily mean that there are exactly 3 men and 5 women in the club, but it does mean that for *every group of 3 men, there is a corresponding group of 5 women.*

Take for example, 30 and 10. You could ask, "What do I have to *multiply* 10 by in order to get 30?"

A ratio is 30 *divided by* 10, which is $\frac{3}{1}$. A ratio is also 30 miles divided by 10 miles, which is also 3, because the units (miles) cancel just like the factor of 10 from $\frac{30}{10}$ to $\frac{3}{1}$.

What if you compare 30 miles to 10 gallons? When you write 30 miles divided by 10 gallons, the units don't cancel. When you write 30 miles divided by 10 gallons, you have four elements: the 30, the unit miles, the 10, and the unit gallons. The 30 and the 10 do cancel to become 3, but the miles and the gallons fail to cancel. You are left with 3 miles per gallon (excellent mileage—for a tank).

In a ratio, the units always cancel. A rate never has uncancelled units. A rate is a pure number, a quantity that is sometimes called a "dimensionless number." For example, 30 feet divided by 10 feet is 3. Not 3 feet, or 3 seconds, or 3 gallons—just 3!

If the comparison of two quantities doesn't succeed in having all its units cancelled, you don't call it a ratio—you call it a rate.

How to Express a Ratio or Rate

A ratio can be expressed in several ways:

- using "**to**" (3 to 5)
- using "**out of**" (3 out of 5)
- using a **colon** (3:5)
- as a **fraction** ($\frac{3}{5}$)
- as a **decimal** (0.6)

All of these expressions represent the same ratio.

INSIDE TRACK

LIKE A FRACTION, a ratio should always be reduced to lowest terms. For example, the ratio of 6 to 10 should be reduced to 3 to 5 (because the fraction $\frac{6}{10}$ reduces to $\frac{3}{5}$).

Here are some examples of ratios in familiar contexts:

The student-teacher ratio at Clarksdale High School is 7 to 1. That means for every 7 students in the school, there is 1 teacher. For example, if Clarksdale has 140 students, then it has 20 teachers. (There are 20 groups, each with 7 students and 1 teacher.)

The baseball team won 27 games and lost 18, for a ratio of 3 wins to 2 losses. Their *win rate* was 60% because they won 60% of the games they played.

Ratios and Totals

A ratio usually tells you something about the *total* number of things being compared. In our first ratio example of a club with 3 men for every 5 women, the club's total membership is a *multiple of* 8 because each group contains 3 men and 5 women.

The following example illustrates some of the *total* questions you could be asked about a particular ratio:

Example

Wyatt bought a total of 12 books, purchasing two $5 books for every $8 book.

- How many $5 books did he buy?
- How many $8 books did he buy?
- How much money did he spend in total?

The total number of books Wyatt bought is a multiple of 3 (each group of books contained **two** $5 books **plus one** $8 book). Because he bought a total of 12 books, he bought 4 groups of books (4 groups × 3 books = 12 books in total).

Total books: **8 $5 books + 4 $8 books** = 12 books

Total cost: $40 + $32 = **$72**

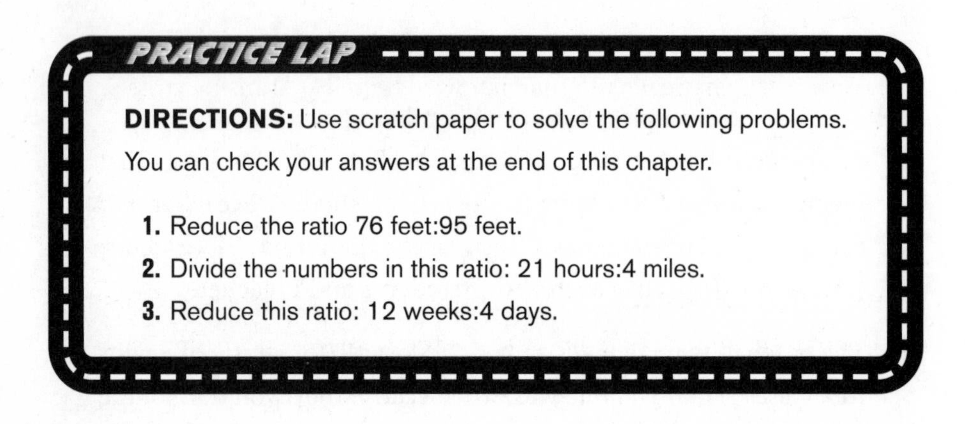

PRACTICE LAP

DIRECTIONS: Use scratch paper to solve the following problems. You can check your answers at the end of this chapter.

1. Reduce the ratio 76 feet:95 feet.
2. Divide the numbers in this ratio: 21 hours:4 miles.
3. Reduce this ratio: 12 weeks:4 days.

PROPORTIONS

A **proportion** states that two ratios are equal to each other. For example, have you ever heard someone say something like this?

Nine out of ten professional athletes suffer at least one injury each season.

The words *nine out of ten* are a ratio. They tell you that $\frac{9}{10}$ of professional athletes suffer at least one injury each season. But there are more than 10 professional athletes. Suppose that there are 100 professional athletes. Then, $\frac{9}{10}$ of the 100 athletes, or 90 out of 100 professional athletes, suffer at least one injury per season. The two ratios are equivalent and form a proportion:

$$\frac{9}{10} = \frac{90}{100}$$

Notice that a proportion reflects equivalent fractions: Each fraction reduces to the same value.

Working with Proportions

The basic math expression for studying proportions is this:

$$\frac{a}{b} = \frac{c}{d}$$

In a typical problem, three of the four quantities are given, and it is up to you to solve the proportion by finding the fourth quantity.

The key mathematical technique involved here is to transform the equation by multiplying both sides by *bd*, the least common denominator:

$$bd(\frac{a}{b}) = bd(\frac{c}{d})$$

After cancelling the *b* on the left-hand side and cancelling the *d* on the right-hand side, you have $ad = bc$.

You can solve for either variable by cross multiplying:

$$a = \frac{bc}{d},\ b = \frac{ad}{c},\ c = \frac{ad}{b},\ \text{and}\ d = \frac{bc}{a}$$

PRACTICE LAP

DIRECTIONS: Use scratch paper to solve the following problems. You can check your answers at the end of this chapter.

4. 8:17 = x:170. Solve this proportion for x.

5. One printer produces 10 pages in 1 minute. How many pages will 5 printers produce in 10 minutes?

6. Ten machines produce 1 carton of books in 3 hours. How long will it take 20 machines to produce 1 carton of books?

MEASURES OF CENTRAL TENDENCY

When dealing with sets of numbers, there are measures used to describe the set as a whole. These are called **measures of central tendency**, and they include mean, median, mode, and range.

Means–No Maybes!

If you have three numbers, say 5, 6, and 7, they have a definite **average**—in this case, 6. It is not *maybe* 6, it is not 6 *with a 95% confidence*, there is not a *large probability* that the average is 6. It is simpler than that. The average of 5, 6, and 7 IS 6. The mean is also known as the average.

The **mean** of a set of values is the sum of the values divided by the number of values. Let's use Abel's and Baker's test scores as examples of calculating the mean:

1	2	3	4	5
Name	**Test scores (in percent)**	**Sum**	***n* (number of test scores)**	**Average or mean**
		sum of column 2	number of numbers in column 2	column 3 divided by column 4
Abel	80, 80, 80, 90, 100	430	5	86
Baker	70, 80, 90, 100, 100	440	5	88

INSIDE TRACK

If you are asked to find the mean of a set of numbers, and the set is evenly spaced apart, such as 2, 4, 6, 8, 10, 12, 14, the mean is the middle number in this set, because there is an odd number of data items. In this example, the mean is 8. If there is an even number of data items, there are two middle numbers; for example, 4, 8, 12, 16, 20, and 24. In this case, the mean is the average of the two middle numbers; 12 + 16 = 28; 28 divided by 2 is 14.

Median–Right in the Middle of Things

The **median** is the value in the middle. You arrange the data in order, and then pick the value in the middle.

For Abel, put aside the lower two scores, which are 80 and 80, and put aside the two larger scores, which are 90 and 100. The score in the middle is 80. Abel's median test score is 80%.

For Baker, put aside the two lower scores, which are 70 and 80, and put aside the two higher scores, which are both 100s. The remaining score is the value in the middle of the set of values. Baker's median test score is 90%.

What if there is an even number of values, like in this data set?

8; 10; 10; 12; 14; 19; 2,008; 1,000,000

There is really no middle value. What to do? In these situations, take the two values nearest the middle. Those would be the fourth and fifth values (after the values have been put in order)—12 and 14. Take the average of these two values. That would be $\frac{(12+14)}{2} = \frac{26}{2} = 13$. The median in this value is 13.

A la Mode

Every set of values has a mean and a median. However, they don't all have a mode.

The **mode** is the value that occurs the largest number of times. For Abel, the mode is 80, because that score occurred three times, more than any other score. For Baker, the mode is 100, because that score occurred twice, more times than any other score occurred for him. Sometimes, the numbers occur the same number of times, so there is no mode or there can be more than one mode.

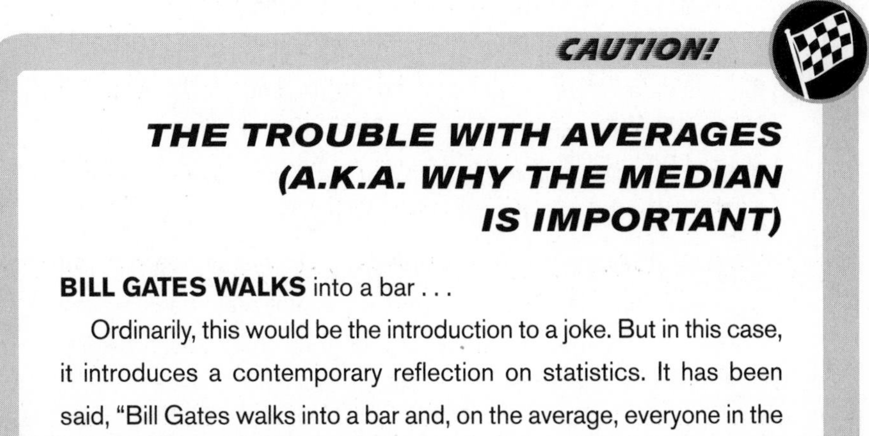

CAUTION!

THE TROUBLE WITH AVERAGES (A.K.A. WHY THE MEDIAN IS IMPORTANT)

BILL GATES WALKS into a bar . . .

Ordinarily, this would be the introduction to a joke. But in this case, it introduces a contemporary reflection on statistics. It has been said, "Bill Gates walks into a bar and, on the average, everyone in the bar becomes a multimillionaire."

What is meant is not exactly that everyone in the bar has actually become a multimillionaire, but rather that the average net financial worth of everyone in the bar is many, many millions of dollars. Bill Gates's financial worth is reportedly several tens of billions of dollars, so if there are 20 people in the bar, even if the 20 people are all broke, their average net worth becomes several tens of billions of dollars divided by 21. The average net worth is many hundreds of millions of dollars.

Notwithstanding the dramatic shift in average net worth, the median shifts only slightly upon Mr. Gates's arrival. And the mode changes not at all.

You see that the average, or mean, can be misleading, and you see that the median can be more robust, so to speak, than the mean. The median can be more *mean*ingful than the mean.

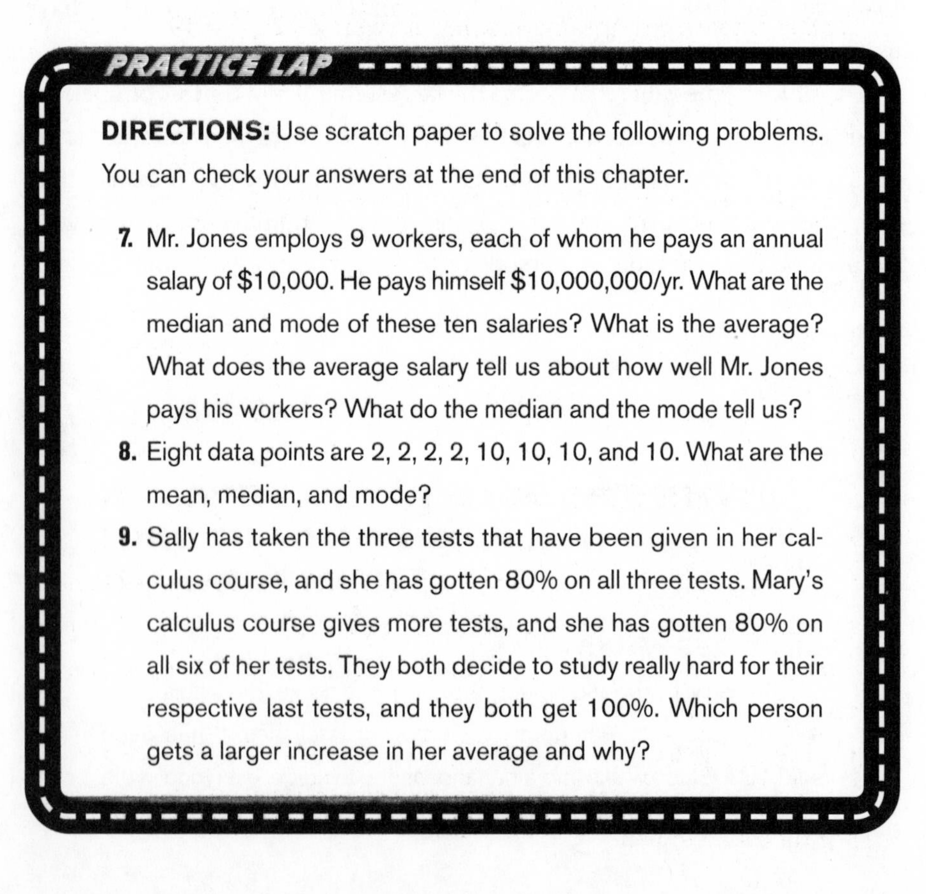

PRACTICE LAP

DIRECTIONS: Use scratch paper to solve the following problems. You can check your answers at the end of this chapter.

7. Mr. Jones employs 9 workers, each of whom he pays an annual salary of $10,000. He pays himself $10,000,000/yr. What are the median and mode of these ten salaries? What is the average? What does the average salary tell us about how well Mr. Jones pays his workers? What do the median and the mode tell us?
8. Eight data points are 2, 2, 2, 2, 10, 10, 10, and 10. What are the mean, median, and mode?
9. Sally has taken the three tests that have been given in her calculus course, and she has gotten 80% on all three tests. Mary's calculus course gives more tests, and she has gotten 80% on all six of her tests. They both decide to study really hard for their respective last tests, and they both get 100%. Which person gets a larger increase in her average and why?

Home on the Range

The **range** indicates how close together the given values are to one another in a set of data.

To calculate the range, find the difference between the two largest and the smallest values in the set of data. Subtract the smallest value from the largest value in the set.

Example

Find the range of ages in the community play, given these ages in years: 68, 54, 49, 40, 39, 39, 24, 22, 20, 10, and 10.

The range of ages is 68 – 10 = 58 years.

ANSWERS

1. Find the prime factors of 76 and of 95.
   Is 76 divisible by 2? Yes. Developing answer: 2: $\frac{76}{2} = 38$.
   Is 38 divisible by 2? Yes. Developing answer: $2 \cdot 2$: $\frac{38}{2} = 19$.
   19 is prime? Yes. Final answer: $76 = 2 \cdot 2 \cdot 19$.
   Is 95 divisible by 2? No.
   Is 95 divisible by 3? No.
   Is 95 divisible by 5? Yes. Developing answer: 5: $\frac{95}{5} = 19$.
   19 is prime? Yes. Final answer: $95 = 5 \cdot 19$.
   Reduce $\frac{76}{95} = \frac{(2 \cdot 2 \cdot 19)}{(5 \cdot 19)}$
   Now, cancel 19 $= \frac{(2 \cdot 2)}{5}$
   Multiply $= \frac{4}{5}$
   76:95 = 4:5
2. $\frac{21 \text{ hours}}{4 \text{ miles}} = (\frac{21}{4})$ hours per mile
   = 5.25 hours per mile
3. $\frac{12 \text{ weeks}}{4 \text{ days}} = \frac{12 \cdot \text{weeks}}{(4 \cdot \text{days})}$. The number multiplies the unit and the parentheses are added to avoid confusion: $\frac{12 \cdot 7 \cdot \text{days}}{(4 \cdot \text{days})}$; substitute 7 days for 1 week. Multiply and divide: $\frac{84}{4} = 21$. The units (days) cancel; 12 weeks is to 4 days as 21 is to 1.
4. $x = 80$; 170 is ten times 17, and 80 is ten times 8.
5. This problem needs two successive proportions. First, let's stay with one printer for a moment.

For the 1 printer:
10 pages:1 minute = x pages:10 minutes.
$x = 100$ pages
Now keeping the ten-minute period:
100 pages:1 printer = y pages:5 printers.
$y = 500$ pages

6. There is an inverse proportion between the number of machines and the time required. Twice as many machines will do the job in half the time. It will take $1\frac{1}{2}$ hours.
7. The median is $10,000/year and the mode is $10,000/year. The average is $10,090,000/year divided by 10, which is $1,009,000/year. The average annual salary is more than a million dollars, but that doesn't tell us much about the nine workers. On the other hand, the median and the mode, in this case, tell us exactly what the workers are being paid.
8. The mean and the median are both 6. Strictly speaking, there is no value that appears the largest number of times (the 2 appears four times and the 10 appears four times), but you can make the interpretation that there are two modes, 2 and 10. A set of numbers like the one in this exercise is said to be *bimodal*, because it has two modes.
9. Mary's average is harder to move, because she has taken more tests than Sally. Sally will have the greater increment in her average. You can prove this by running the numbers. They both start with an average of 80%. Sally's average becomes $\frac{340\%}{4} = 85\%$, for an increment of 5%. Mary's average becomes $\frac{580\%}{7} = 82.86\%$, for an increment of only 2.86%.

11 Posttest

If you have completed all the chapters in this book, then you are ready to take the posttest to measure your progress. The posttest has 50 questions covering the topics you studied in this book. Although the format of the posttest is similar to that of the pretest, the questions are different.

Take as much time as you need to complete the posttest. When you are finished, check your answers with the answer key at the end of the posttest. Along with each answer is a number that tells you which chapter of this book teaches you the math skills needed for that question. Once you know your score on the posttest, compare the results with the pretest. If you scored better on the posttest than on the pretest, congratulations! You have profited from the hard work. At this point, you should look at the questions you missed, if any. Do you know why you missed the question, or do you need to go back to the chapter and review the concept?

If your score on the posttest doesn't show much improvement, take a second look at the questions you missed. Did you miss a question because of an error you made? If you can figure out why you missed the question, then you understand the concept and just need to concentrate more on accuracy when taking a test. If you missed a question because you did not know how to work a problem, go back to the chapter and spend time working that type of problem. Take time to understand basic math and pre-algebra thoroughly. You need a solid foundation in basic algebra if you plan to use this information or progress to a higher level of algebra. Whatever your score on this posttest, keep this book for review and future reference.

1.

$$\begin{array}{r} 284 \\ 119 \\ 243 \\ +\ 745 \\ \hline \end{array}$$

2. What is the place value of the 4 in the number 54,321?

3. List the five smallest prime numbers.

4. What are the prime factors of 60?

5. Calculate 23×45.

6. Calculate $\frac{2}{3} + \frac{3}{4}$. Write your answer as a mixed number.

7. $\frac{2}{3} \cdot \frac{3}{4} =$

8. $\frac{2}{3} \div \frac{3}{4} =$

9. Calculate $\frac{2}{3} - \frac{1}{4}$.

10. Calculate $5\frac{1}{4} - 3\frac{2}{3}$.

11. Convert 0.45% to a decimal.

12. Express 2.3 as a percent.

13. In a bag of 20 chocolates, 16 are crème-filled. What percent of the chocolates are crème-filled?

14. What percent of 5 is 4?

15. Calculate $4 - -8$.

16. $5 - 3 + 2 + 8 - 1 =$

17. How much is $-8 \div -4$?

18. Express $-(\frac{-4}{-5})$ as a decimal.

19. Using these categories of numbers—whole numbers, integers, rational numbers, and real numbers—the following numbers belong to which categories?

(a) 15

(b) $\frac{2}{3}$

(c) $\sqrt{2}$

20. How many pennies are q quarters worth?

21. Distribute q in the expression $q(a + b + c)$.

22. Evaluate $100a + 10b + c$ when $a = 2$, $b = 3$, and $c = 4$.

23. Solve $3x + 6 = 48$ for x.

24. Which is larger, -8 or the reciprocal of -8?

25. Express $\frac{2}{5}$ as a product.

26. Does every number have a reciprocal?

27. Simplify $\frac{a^2b^3c^4d^5}{(ab^3cd^3e^2)}$.

28. Marcel needs 250 cookies for a birthday party. He will make oatmeal raisin, macaroons, and chocolate chunk cookies. He wants 25% of the cookies to be oatmeal raisin and 15% of the cookies to be macaroons. How many chocolate chunk cookies must Marcel bake?

29. Henrietta leaves $10,000 in the bank accumulating 5% annual interest, compounded annually. How much does she have in her account at the end of 3 years?

30. A video game is marked up 30%. A week later, it is marked down 30%. What is the net percent increase or decrease?

31. A coat is marked up 20% and then, on the next day, marked up 30%. What is the net effect of the two markups?

32. Which is larger, 0.0023 or 0.023?

33. 3,279 + 32.79 + 3.279 =

34. $-|-5| =$

35. How many cubic centimeters are in 3 cubic feet? (Remember: 1 yard = 3 feet and 1 foot = 12 inches.)

36. How many cubic inches in 1 cubic yard?

37. Find the perimeter of a square whose side is 12 centimeters.

38. Find the area of a triangle with base 5 units and height 8 units.

39. How long is 5 hours, 10 minutes, 34 seconds plus 7 hours, 55 minutes, and 43 seconds?

40. The circumference of a circle is 36π. Find the radius of the circle.

41. Reduce the ratio 20:15.

42. John uses 1.2 cups of sugar for each apple pie he makes. How many cups of sugar will he use to make 5 apple pies?

43. Holly can complete a certain job in 12 days. With the help of her two brothers, Sam and Fred, who each work at the same rate Holly does, how long will it take them to complete the job?

44. A big circle has three times the radius of a small circle. How many times larger is the area of the big circle, compared to the area of the small circle?

45. What is the volume of a cylinder that has a radius of 10 feet and a height of 20 feet?

46. What is the volume of a sphere that has a radius of 1 foot?

47. What is the perimeter of a rectangle that has a length of 8 inches and a width of 2 feet 5 inches?

48. What is the (a) mean, (b) median, (c) mode, and (d) range of the following set of data?

5, 6, 8, 8, 6, 6, 8, 5, 8, 8, 8, 5, 6, 8

49. If 5 is added to the data set in question 48, which of the following would change and which would stay the same: mean, median, mode, and range?

50. What is the (a) mean, (b) median, (c) mode, and (d) range for the following set of data?

50, 80, 80, 80, 50, 60, 80, 50, 60, 80, 80, 60, 60, 80

ANSWERS

1. These numbers add to 1,391. For more help with this concept, see Chapter 3.
2. The 4 in 54,321 is in the thousands place. For more help with this concept, see Chapter 3.
3. The five smallest prime numbers are 2, 3, 5, 7, and 11. For more help with this concept, see Chapter 3.
4. Is 60 divisible by 2? Yes. Developing answer: 2: $\frac{60}{2} = 30$.
Is 30 divisible by 2? Yes. Developing answer: $2 \cdot 2$: $\frac{30}{2} = 15$.

Is 15 divisible by 2? No. Is 15 divisible by 3? Yes. Developing answer: $2 \cdot 2 \cdot 3. \frac{15}{3} = 5$.

Final answer: $60 = 2 \cdot 2 \cdot 3 \cdot 5 = 2^2 \cdot 3 \cdot 5$.

For more help with this concept, see Chapter 3.

5.

$$\begin{array}{r} 23 \\ \times\, 45 \\ \hline 115 \\ 92 \\ \hline 1{,}035 \end{array}$$

For more help with this concept, see Chapter 3.

6. $\frac{2}{3} + \frac{3}{4} = \frac{8}{12} + \frac{9}{12} = \frac{(8+9)}{12} = \frac{17}{12} = 1\frac{5}{12}$. For more help with this concept, see Chapter 4.

7. $\frac{2}{3} \cdot \frac{3}{4} = \frac{2}{4} = \frac{1}{2}$. You cancelled out the 3s and then divided the numerator and denominator by 2 in order to reduce the fraction. For more help with this concept, see Chapter 4.

8. $\frac{2}{3} \div \frac{3}{4} = \frac{2}{3} \cdot \frac{4}{3} = \frac{8}{9}$. Multiply by the reciprocal of $\frac{3}{4}$, then multiply the numerators and the denominators. For more help with this concept, see Chapter 4.

9. $\frac{2}{3} - \frac{1}{4} = \frac{8}{12} - \frac{3}{12} = \frac{(8-3)}{12} = \frac{5}{12}$. For more help with this concept, see Chapter 4.

10. $5\frac{1}{4} - 3\frac{2}{3} = 5\frac{3}{12} - 3\frac{8}{12} = 4\frac{15}{12} - 3\frac{8}{12} = 1\frac{7}{12}$. You borrowed from the 5 to make $\frac{12}{12}$. Then, you subtracted the whole numbers and the fractions. For more help with this concept, see Chapter 4.

11. $0.45\% = 0.45 \cdot 0.01\% = 0.0045$. For more help with this concept, see Chapter 6.

12. $2.3 \cdot 100\% = 230\%$. For more help with this concept, see Chapter 6.

13. The fraction of crème-filled chocolates is $\frac{16}{20}$. This equals 0.8, or 80%. Eighty percent of the chocolates are crème-filled. For more help with this concept, see Chapter 6.

14. Change the question to "What fraction of 5 is 4?" Four is $\frac{4}{5}$ of 5. The percent equivalent of $\frac{4}{5}$ is 80%. For more help with this concept, see Chapter 6.

15. $4 - -8 = 4 + 8 = 12$. For more help with this concept, see Chapter 7.

16. $5 - 3 + 2 + 8 - 1 = 5 + -3 + 2 + 8 + -1 = 15 + -4 = 11$. For more help with this concept, see Chapter 7.

17. A negative number divided by a negative number gives a positive answer, so $8 \div 4 = 2$. For more help with this concept, see Chapter 7.

18. $-(\frac{-4}{-5}) = -(\frac{4}{5}) = -0.8$. For more help with this concept, see Chapters 4 and 7.

19. (a) 15 belongs to all the categories.
(b) $\frac{2}{3}$ is a fraction. It belongs to the rational and the real number categories.
(c) $\sqrt{2}$ is irrational. It belongs only to the real number category.
For more help with this concept, see Chapter 3.

20. q quarters are worth $25q$ pennies. For more help with this concept, see Chapter 8.

21. $q(a + b + c) = qa + qb + qc$. For more help with this concept, see Chapter 8.

22. $100a + 10b + c = (100 \cdot 2) + (10 \cdot 3) + (4) = 200 + 30 + 4 = 234$. For more help with this concept, see Chapter 8.

23. $3x + 6 = 48$
$3x = 42$; Add -6 to both sides.
$x = 14$; Divide both sides by 3.
Check the answer by plugging 14 into the original equation:
$3(14) + 6 = 48$
$42 + 6 = 48$
$48 = 48$
For more help with this concept, see Chapter 8.

24. Compare -8 to the reciprocal of -8, which is $-\frac{1}{8}$: $-\frac{1}{8}$ is closer to the origin than -8, which means it is to the right of -8 on the number line; $-\frac{1}{8}$ is larger than -8. For more help with this concept, see Chapters 4 and 7.

25. $\frac{2}{5} = 2$ times the reciprocal of 5: $\frac{2}{5} = 2(\frac{1}{5}) = 2(0.2) = 0.4$. For more help with this concept, see Chapter 4.

26. Zero doesn't have a reciprocal. Zero times any number is zero, so there is no number you can multiply zero by to get 1. Every non-zero number has a reciprocal. For more help with this concept, see Chapter 4.

27. $\frac{a^2b^3c^4d^5}{(ab^3cd^3e^2)} = \frac{ac^3d^2}{e^2}$. Cancel out the variables that can be cancelled out: $\frac{a^2}{a} = a$; $\frac{b^3}{b^3} = 1$; $\frac{c^4}{c} = c^3$; and $\frac{d^5}{d^3} = d^2$. For more help with this concept, see Chapter 8.

28. Marcel needs 150 chocolate chunk cookies for the party; 60% of the cookies should be chocolate chunk, and 60% of 250 is 150. For more help with this concept, see Chapter 6.

29. At the end of the first year, Henrietta's bank account stands at \$10,000(1.05). At the end of 2 years, her account has been multiplied by 1.05 again, and stands at $[\$10{,}000(1.05)] \times 1.05 = \$10{,}000(1.05)^2$. At the end of the third year, the account has $\$10{,}000(1.05)^3 = \$10{,}000(1.157625) = \$11{,}576.25$. For more help with this concept, see Chapter 8.

30. If the starting price is P, after the markup it is $1.3P$. Then, it is marked down 30%. Now, the price is $0.7(1.3P) = 0.91P$. The net result is a markdown of 9% of the original price. For more help with this concept, see Chapter 8.

31. If the original price is P, then after the first markup the price is $1.2P$. Then, the cost is marked up again, this time by 30%. The final price is $1.3(1.2P) = 1.56P$, representing an effective markup of 56% of the original price. For more help with this concept, see Chapter 8.

32. 0.023 is larger. To determine the answer to this question, you should have compared decimal places. Both numbers have 0 in the tenth place; however, 0.023 has a 2 in the hundredths place, while 0.0023 has a 0. For more help with this concept, see Chapter 5.

33. Turn on the decimal point in 3,279, line up the numbers, pad some zeros, and do the addition without regard to the decimal point column, except to copy the decimal point.

$$\begin{array}{r} 3{,}279.000 \\ 32.790 \\ +\quad 3.279 \\ \hline 3{,}315.069 \end{array}$$

For more help with this concept, see Chapter 5.

34. $-|-5| = -5$; $|-5| = 5$, and you want the opposite of this value. For more help with this concept, see Chapter 7.

35. 3 ft.^{3} = 3 · ft.^{3} (The number multiplies the unit.)
= 3 · 28.32 cm^3 (Substitute for ft.^{3} from the conversion tables.)
= 84.96 cm^3 (Do the multiplication.)
For more help with this concept, see Chapter 9.

36. 1 yd.^{3} = (36 in.)^3 (Substitute 36 inches for 1 yard.)
= 36 · in. · 36 · in. · 36 · in. (Write out the cube.)
= 36^3 · in.^{3} (Rearrange the factors.)
= 46,656 in.^{3} (Do the multiplication.)
For more help with this concept, see Chapter 9.

37. Add up the length of all four sides: 12 + 12 + 12 + 12 = 48 centimeters. For more help with this concept, see Chapter 9.

38. The area formula for a triangle is $\frac{1}{2}bh$, so $\frac{1}{2}(5)(8) = 20$ units. For more help with this concept, see Chapter 9.

39. Adding the hours, minutes, and seconds separately, you have 12 hours, 65 minutes, and 77 seconds. Convert the 77 seconds to 1 minute and 17 seconds, so that we now have 12 hours, 66 minutes, and 17 seconds. Now, convert the 66 minutes to 1 hour and 6 minutes. The result is 13 hours, 6 minutes, and 17 seconds. For more help with this concept, see Chapter 9.

40. The diameter of the circle is 36, so the radius would be half this length, or 18. For more help with this concept, see Chapter 9.

41. Divide both elements of the ratio by 5, so 20:15 = 4:3. For more help with this concept, see Chapter 10.

42. In this problem, 1.2 cups of sugar is to 1 apple pie as x cups of sugar is to 5 apple pies. This translates to $\frac{1.2}{1} = \frac{x}{5}$. To solve for x, cross multiply: $x = (1.2)(5) = 6$. John will need 6 cups of sugar for the 5 apple pies. For more help with this concept, see Chapter 10.

43. Holly does the job in 12 days. Three people require how many days? This is an inverse proportion, because more people will need less time. Three times as many people will require $\frac{1}{3}$ the time. They will require $(\frac{1}{3})12$ days = 4 days. For more help with this concept, see Chapter 10.

44. The area of the larger circle is 9 times the area of the smaller circle. The formula for area is $A = \pi r \times r$. When you replace r with $3r$, you get $A = \pi(3r)(3r) = 9(\pi r^2)$. For more help with this concept, see Chapter 9.

45. The formula for the volume of a cylinder is $V = \pi r^2 h$. For $r = 10$ feet and $h = 20$ feet, the volume is $V = \pi 100$ ft.^{2} · 20 feet = $\pi 2{,}000$ ft.^{3} = 6,280 ft.^{3} For more help with this concept, see Chapter 9.

46. The formula for the volume of a sphere is $V = (\frac{4}{3})\pi r^3$. Plugging in 1 foot for r, you get $V = (\frac{4}{3})\pi \cdot$ ft.^{3} = 4.2 ft.^{3}. For more help with this concept, see Chapter 9.

47. The formula for perimeter is $P = 2l + 2w$. Plugging in the given values for l (the length) and w (the width), we have P = 2(8 inches) + 2(2 feet + 5 inches) = 16 inches + 4 feet + 10 inches = 4 feet + 26 inches. Convert the 26 inches to 2 feet and 2 inches. The result is that the perimeter is 6 feet + 2 inches. For more help with this concept, see Chapter 9.

48. The sum in this set of numbers is 95. (a) The mean is $\frac{95}{14} = 6.79$. (b) The median is the average of the seventh value and the eighth value. Those values are 6 and 8, so the median is 7. (c) The mode is the value that occurs the most frequently, so the mode is 8, which occurs 7 times. (d) The range is the largest number (8) minus the smallest number (5). The range is 8 – 5 = 3. For more help with this concept, see Chapter 10.

49. The mean, median, and mode would each increase by 5. The range would not change. For more help with this concept, see Chapter 10.

50. (a) The mean is $\frac{950}{14}$, or approximately 67.86. (b) The median is the average of the seventh number and the eighth number. Those numbers are 60 and 80, so the median is 70. (c) The mode is the number that occurs the most frequently, so the mode is 80, which occurs 7 times. (d) The range is the largest number (80) minus the smallest number (50). The range is 80 – 50 = 30. For more help with this concept, see Chapter 10.

Glossary

Absolute value: For a number, this is the distance, or number of units, from the origin on a number line. Absolute value is the size of the number and is always positive.

Area: A measure of how many square units it takes to cover a closed figure.

Circle: The set of all points equidistant from one given point, called the *center.* The center point defines the circle, but it is not on the circle.

Decimal: Numbers related to or based on the number ten. The place value system is a decimal system because the place values (units, tens, hundreds, etc.) are multiples of ten.

Denominator: The bottom number in a fraction. *Example:* 2 is the denominator in $\frac{1}{2}$.

Diameter: A line segment that passes through the center of the circle whose endpoints are on the circle. The diameter is twice the radius: $d = 2r$.

Difference: The result of subtracting one number from the other.

Divisible by: A number is divisible by a second number if that second number divides *evenly* into the original number. *Example:* 10 is divisible by 5 ($10 \div 5 = 2$, with no remainder). However, 10 is not divisible by 3. (See *Multiple of*)

Even integer: Integers that are evenly divisible by 2, such as –4, –2, 0, 2, 4, and so on. (See *Integer*)

Fraction: The result of dividing two numbers. When you divide 3 by 5, you get $\frac{3}{5}$, which equals 0.6. A fraction is a way of expressing a number that involves dividing a top number (the numerator) by a bottom number (the denominator).

Improper fraction: A fraction whose numerator is greater than or equal to its denominator.

Integer: A number along the number line, such as –3, –2, –1, 0, 1, 2, 3, and so on. Integers include whole numbers and their negatives. (See *Whole number*)

Irrational number: A real number that is not rational—that is, an irrational number cannot be expressed as the ratio of integers.

Mean: The average of a set of data.

Median: The middle value in a set of numbers that are arranged in increasing or decreasing order. If there are two middle numbers, it is the average of these two.

Mixed number: A number with an integer part and a fractional part. Mixed number can be converted into improper fractions.

Mode: The value in a set of numbers that occurs most often. There can be one mode, several modes, or no mode.

Multiple of: A number is a multiple of a second number if that second number can be multiplied by an integer to get the original number. *Example:* 10 is a multiple of 5 ($10 = 5 \times 2$); however, 10 is not a multiple of 3. (See *Divisible by*)

Negative number: A number that is less than zero, such as –1, –18.6, or –14.

Numerator: The top part of a fraction. *Example:* 1 is the numerator in $\frac{1}{2}$.

Odd integer: An integer that is not evenly divisible by 2, such as –5, –3, –1, 1, 3, and so on.

Polygon: A closed-plane figure made up of line segments.

Positive number: A number that is greater than zero, such as 2, 42, $\frac{1}{2}$, 4.63.

Prime number: An integer that is divisible only by 1 and itself, such as 2, 3, 5, 7, 11, and so on. All prime numbers are odd, except for 2. The number 1 is not considered prime.

Product: The result of a multiplication problem.

Proper fraction: A fraction whose numerator is less than its denominator.

Proportion: An equation that states that two ratios are equal.

Quadrilateral: A polygon with four sides.

Quotient: The answer you get when you divide. *Example:* 10 divided by 5 is 2; the quotient is 2.

Percent: A ratio that compares numerical data to one hundred. The symbol for percent is %.

Perimeter: The measure around a polygon.

Proper fraction: A fraction whose numerator is less than its denominator.

Radius: The line segment whose one endpoint is at the center of the circle and whose other endpoint is on the circle. The radius is one-half the length of the diameter; $r = \frac{1}{2}(d)$.

Range: A number that indicates how close together the values are to one another in a set of data.

Rate: A ratio comparing two items with unlike units.

Ratio: A comparison of two things using numbers.

Rational number: A number that can be expressed as the quotient of an integer divided by a non-zero integer.

Real numbers: All the numbers on the number line.

Reciprocal: The multiplicative inverse of a fraction. Any number multiplied by its reciprocal equals 1. For example, $\frac{2}{1}$ is the reciprocal of $\frac{1}{2}$.

Remainder: The number left over after division. *Example:* 11 divided by 2 is 5, with a remainder of 1.

Sum: The result of adding two or more numbers.

Triangle: A polygon with three sides.

Volume: A measurement of how many cubic units it takes to fill a solid figure.

Whole number: Numbers you can count on your fingers, such as 1, 2, 3, and so on. All whole numbers are positive.